THE ELEMENTS OF
PALAEONTOLOGY

THE ELEMENTS OF PALAEONTOLOGY

Rhona M. Black

CAMBRIDGE UNIVERSITY PRESS

CAMBRIDGE

LONDON · NEW YORK · MELBOURNE

Published by the Press Syndicate of the University of Cambridge
The Pitt Building, Trumpington Street, Cambridge CB2 1RP
32 East 57th Street, New York, NY 10022, USA
296 Beaconsfield Parade, Middle Park, Melbourne 3206, Australia

Library of Congress catalogue card number: 78-149442

ISBN 0 521 07445 2 hard covers
ISBN 0 521 09615 4 paperback

First published 1970
Reprinted 1972
Reprinted with additions 1973
Reprinted 1975, 1978, 1979

First printed in Great Britain by
Jarrold and Sons Ltd, Norwich
Reprinted in Great Britain
at the Alden Press, Oxford

CONTENTS

CONTENTS

PREFACE

This book is written primarily for the use of students who require a knowledge of a wide field in palaeontology at an introductory level. All the major groups of invertebrate fossils are dealt with, and an outline of the palaeontology of the vertebrates and plants is included. Some emphasis is given to biological aspects of palaeontology since it is becoming increasingly clear that the interpretation of fossils and fossil assemblages, from a biological point of view, throws much light on wider geological problems in sedimentation, stratigraphy, and palaeogeography.

NOTE ON THE THIRD IMPRESSION

The need for a further reprint has provided an opportunity to revise figure 1 and to state geological ages directly in legends as a help to those unfamiliar with British rock formations.

1973

ACKNOWLEDGEMENTS

The author is indebted to Mr Tony Cartwright and to Dr B.M. Funnell for their comments on the draft of preliminary chapters, to Miss Dianne Edwards for her comments on the section on plants and to Dr T. Kemp for his comments on the section on vertebrates. Thanks are due, too, to Mr A.G. Brighton, and Professor H.B. Whittington for facilities. I am most grateful to Professor O.M.B. Bulman for various amendments of the typescript which he suggested, and to Dr W.W. Black whose constructive criticism of the typescript has greatly improved this book.

Grateful acknowledgement is made to the following who helped by redrawing diagrams: Mr D. Batten (figs. 32, 35, 37, 38, 194, 197); Mrs J. Friend (figs. 92, 94, 143); Dr R.B. Rickards (figs. 129, 130, 132), Dr T. Kemp (figs. 147, 150, 155, 160), and Dr J.K. Ingham (figs. 112, 116, 119) who also included some of his own drawings in the trilobite diagrams, notably fig. 112c and e.

Diagrams of fossils have been drawn from specimens where possible; otherwise they are modified after illustrations in monographs of the Palaeontographical Society and various palaeontological journals. Skeletal details in sketches of fossil vertebrates are modified after Zittel (*Textbook of Palaeontology*) and (with permission) A.S. Romer (*Vertebrate Palaeontology*).

Many of the photographs are of specimens in the Sedgwick Museum or loaned by friends and are the work of Dr C.P. Hughes, Mr A. Barlow and Mr D. Bursill, whose help I acknowledge with especial gratitude. I am much indebted to Dr W.D.I. Rolfe for his help in selecting the photographs, some specially prepared, of exhibits in the Hunterian Museum and to the Hunterian Museum for supplying these photographs: figs. 4, 41, 56, 115, 123, 124, 154, 157, 158, 166, 171, 173, 177, 196; also fig. 117, by courtesy of Dr J.K. Ingham; figs. 16, 43, 84, 95a–c, 133 and 137, by courtesy of Dr W.D.I. Rolfe; and fig. 165, by courtesy of Norman Bros. My sincere thanks are also extended to those who made available the following photographs: Mr A.M. Honeyman (fig. 5); The National Institute of Oceanography (fig. 12); Dr R.G. West (fig. 14; a and c reproduced from West, *Pleistocene Geology and Biology*, with permission from the Longman Group Ltd); Dr P.M. Kier (figs. 62, 68, 69, 85, 106); Professor H.B. Whittington (figs. 113, 114, Plate III); Professor F.M. Carpenter (figs. 125, 126, Plate I); Dr G.P. Larwood (fig. 141); Dr C.P. Hughes and Professor H.B. Whittington (fig. 142); Dr D. Gobbett (fig. 144a); Professor A. Heintz (figs. 148b, 149); The British Museum (Natural History) (figs. 148a, 156, 162, 163, 164, 167, 168, 169, 170, 176, 199); Dr P. Echlin and the Cambridge Scientific Instrument Co. Ltd (fig. 190b); Dr M. Black (figs. 189, 190a, 191); Dr T.D. Ford (fig. 202, by permission of the Council of the Yorkshire Geological Society).

PART ONE

Introduction

1

Introduction

Palaeontology is the study of prehistoric animals and plants of which remains or other indications are found in sedimentary rocks, and which are described as FOSSILS.

FOSSILS. Organic origin is implicit in the term 'fossil'. A fossil represents part of a once living creature or plant, or has been formed by the action of a once living organism. For example, the bones of a dinosaur are fossils, and so also are the footprints which a dinosaur made in wet sand. The term 'fossil' implies great age since it refers only to the remains of organisms which died in prehistoric times. It is not, however, implied that all fossils must represent extinct organisms; a wide variety of forms living today are known also from the fossil record.

RELATIONSHIP OF PALAEONTOLOGY WITH OTHER SUBJECTS. For a full understanding of palaeontology, acquaintance with other scientific subjects, especially the biological sciences, is important. Thus zoology and botany give the information about living organisms without which no interpretation of fossils could be made. Equally, of course, palaeontology provides the biologist with evidence about the ancestry of living animals and plants.

Fossils have an intrinsic value, but their study is a fundamental part of geology. In economic geology, they are important as a means of identifying the rock formations in which fuels like coal and oil occur. Their main importance to the geologist however, lies in aspects of historical geology. This is the branch of geology which aims to interpret the record of events in earth's history, of past geography, climate and local environments, from the stratified rocks in which fossils occur. In this way, palaeontology impinges closely on STRATIGRAPHY.

STRATIGRAPHY deals with the nature and origin of stratified rocks; with their sequence in the earth's crust; and with their correlation, i.e. the identification of isolated outcrops of rocks with those of similar age elsewhere. The first practical step in stratigraphy is the grouping of strata into lithological units, e.g. limestones, clays, etc. The second step is to establish the sequence of these units in time. For this purpose, the basis of all stratigraphical work is the principle of superposition of strata – simply, that in any normal undisturbed sequence of sedimentary rocks, a stratum is younger than the one on which it rests. This principle has been used to establish the correct sequence of strata in the stratigraphical column.

Early work in this field showed that the fossils which occur in any one part of the stratigraphical column are distinctive and different from those which occur at other levels. This made possible the delineation of groups of strata containing particular characteristic fossils and known as ZONES. Once the order of these zones was established it became possible to refer any set of sedimentary rocks to its correct position in the stratigraphical column by examination of the fossil content.

STRATIGRAPHICAL COLUMN. The complete succession of stratified rocks which contain fossils is summarised in the STRATIGRAPHICAL COLUMN in which the strata are combined into major groups and systems and arranged in sequence with the oldest rocks at the base and the youngest at the top (fig. 1). Each system contains a number of zones. The earliest rocks in which fossils occur in appreciable numbers, the Cambrian, overlie still older rocks, the Pre-Cambrian, which cannot be correlated by means of fossils.

CLASSIFICATION OF ORGANISMS. Some method of distinguishing one particular type of organism amongst the multiplicity of diverse forms of life is necessary. For this purpose a system of naming, the BINOMIAL SYSTEM, is used universally, each organism which is a recognisably different type being given a name consisting of two parts, a SPECIFIC name and a GENERIC name. The specific name, always written with a small initial letter, denotes the SPECIES. A species is the smallest unit of division in common use, and as a biological concept it defines a group of similar animals which can breed only within that group. A fossil species must obviously be identified by the characters of the hard parts and typically these show some variation. If sufficient individuals from the same bed are available to show that the variation is continuous, they may be referred to the same species without hesitation. With smaller numbers, however, the only practicable definition of species is 'a collection of individuals which show only minor differences'.

The generic name, always written with a capital initial letter, denotes the GENUS to which the species belongs. A genus is an arbitrary unit consisting of a number of species which have similar features and are closely related.

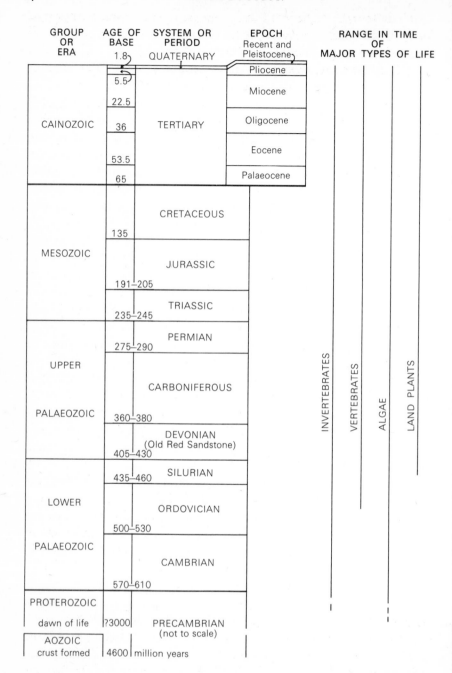

GROUP OR ERA	AGE OF BASE	SYSTEM OR PERIOD	EPOCH	RANGE IN TIME OF MAJOR TYPES OF LIFE
	1.8	QUATERNARY	Recent and Pleistocene	
CAINOZOIC	5.5	TERTIARY	Pliocene	
	22.5		Miocene	
	36		Oligocene	
	53.5		Eocene	
	65		Palaeocene	
MESOZOIC	135	CRETACEOUS		
	191–205	JURASSIC		
	235–245	TRIASSIC		
UPPER PALAEOZOIC	275–290	PERMIAN		
	360–380	CARBONIFEROUS		
	405–430	DEVONIAN (Old Red Sandstone)		
LOWER PALAEOZOIC	435–460	SILURIAN		
	500–530	ORDOVICIAN		
	570–610	CAMBRIAN		
PROTEROZOIC dawn of life	?3000	PRECAMBRIAN (not to scale)		
AOZOIC crust formed	4600	million years		

(Range columns: INVERTEBRATES, VERTEBRATES, ALGAE, LAND PLANTS)

1   Geological time scale.

The scale used for the Cainozoic is increased to include more detail. Where a range of age is given this reflects uncertainty about the precise age of that horizon.

The following are the MAIN categories used in the higher classification of organisms. Just as related species comprise a genus, so related genera make up a FAMILY, related families an ORDER, orders a CLASS, and classes a PHYLUM (animals) or DIVISION (plants). The phylum, or division is the largest category of the two primary groups into which most organisms may be fitted, namely the ANIMAL and PLANT kingdoms; it is an assemblage of organisms having a common structural plan. In cases where additional categories are required the prefixes 'sub-' or 'super-' may be added thus: suborder; superfamily.

Example to illustrate the classification of animals:
Man is classified thus:

Phylum: Chordata
Subphylum: Vertebrata
Class: Mammalia
Order: Primates
Family: Hominidae
Genus: *Homo*
Species: *sapiens*

The binomial system of classification, while comprehensive, is also flexible; it allows for modifications in the light of further information about particular organisms and their ancestry. Inevitably, the system is hedged in by conventions. These ensure, as nearly as possible, precise identification of fossils. When a fossil is first named, a particular specimen is figured and described as the TYPE which (ideally) is lodged in a museum. Thus, if doubt exists about the identity of a particular fossil, it may be compared with the type of a species.

Confusion sometimes arises because a fossil is called by one generic name in an earlier textbook, but by a different generic name in a later textbook. There are several reasons why a change of name may occur. It may simply reflect improved knowledge about the actual relationship of the fossil. Or, a worker who named the fossil may have been unaware that it had already been described under a different name by an earlier worker. Since the first name proposed for a fossil has priority, its subsequent discovery invalidates the later name. Occasionally, too, a fossil is given a name which, unknown to the author, is already in use for another animal, so that when the mistake is discovered a new name must be chosen. Where the invalid name has become familiar in literature it may sometimes be placed within square brackets after the new name, thus: *Ogygiocaris* [*Ogygia*].

PRINCIPAL DIVISIONS OF ANIMALS AND PLANTS OF INTEREST TO PALAEONTOLOGISTS

Animal kingdom

Phylum and geological range	*Examples and summary of characteristics*
PROTOZOA Cambrian to Recent	One-celled animals including foraminifera with calcareous shells, and radiolarians with siliceous shells.
PORIFERA Cambrian to Recent	Sponges; many-celled, sessile, aquatic animals which lack definite tissues and organs; a skeleton of calcareous, siliceous or organic spicules may be present.
COELENTERATA ? Pre-Cambrian to Recent	Corals and jellyfish; simple forms having a body cavity with a single opening. Corals have a calcareous skeleton; jellyfish lack a skeleton and are rare as fossils.
BRYOZOA Ordovician to Recent	'Sea-mosses'; small colonial forms more complexly organised than coelenterates and with the skeleton mainly calcareous.
BRACHIOPODA Cambrian to Recent	Lamp shells; sessile, marine shellfish enclosed by two dissimilar valves which are usually calcareous.
MOLLUSCA Cambrian to Recent	Gastropods, bivalves, ammonites and belemnites; diverse, typically aquatic, mobile animals with a highly organised soft body enclosed by a calcareous shell of one, two or more parts.
ANNELIDA Pre-Cambrian to Recent	Segmented worms which may build calcareous tubes or may leave tracks or burrows.
ARTHROPODA Cambrian to Recent	Trilobites, crustaceans (crabs), insects, arachnids (spiders), eurypterids; segmented animals with jointed limbs and covered by a chitinous skeleton.
ECHINODERMATA Cambrian to Recent	Echinoids (sea urchins), crinoids, starfish and lesser, extinct, groups; exclusively marine with mainly five-rayed symmetry and a water-circulating system used in feeding, breathing and moving; skeleton of calcite plates.

HEMICHORDATA ? Ordovician to Recent	Includes worm-like forms and some colonial forms with a scleroproteinic skeleton.
Upper Cambrian to Lower Carboniferous	Graptolites are included in this phylum; colonial marine forms with a scleroproteinic skeleton.
CHORDATA Ordovician to Recent	Fish, amphibians, reptiles, birds, and mammals, i.e. the vertebrates with an internal bony skeleton, a backbone (or notochord) and gill slits.

Plant kingdom

Division and geological range

ALGAE Pre-Cambrian to Recent	Aquatic one-celled and many-celled plants some of which have calcareous or siliceous skeletons.
BACTERIA Pre-Cambrian to Recent	Simple microscopic plants which lack chlorophyll and live mainly on organic matter.
FUNGI ? Devonian to Recent	Filamentous plants which lack chlorophyll and live on organic matter.
BRYOPHYTES Upper Devonian to Recent	Mosses; simplest of land-growing plants.
PTERIDOPHYTES Silurian to Recent	Psilophytes, ferns, horsetails, club-mosses; vascular plants which reproduce by spores.
GYMNOSPERMS Devonian to Recent	Pteridosperms (i.e. seed ferns), cycadophytes, conifers; primitive seed plants.
ANGIOSPERMS Cretaceous to Recent	Flowering seed plants with the ovules enclosed.

2

Preservation of fossils

Nature of the hard parts of organisms

The hard parts of an organism may take the form of a rigid structure within the body or enclosing it, or of separate units which fall apart when the soft tissues decay after death.

In VERTEBRATES the principal component of the skeleton is BONE. The matrix of bone consists mainly of collagen (a fibrous scleroprotein) hardened by mineral salts, largely calcium phosphate; it has a cellular structure ramified by channels containing blood-vessels. TEETH form a minor part of the skeleton; like bone they also consist of calcium salts, but their structure is denser and they have a surface layer of ENAMEL, almost pure calcium phosphate and carbonate. They are relatively resistant to decay and are commonly fossilised.

In many INVERTEBRATES the body consists only of soft tissue, but those of most interest to palaeontologists have a shell or skeleton which is relatively resistant to decay. In some groups the skeleton consists of a complex organic substance. In insects, for instance, it consists of CHITIN, a fibrous nitrogen-containing polysaccharide (carbohydrate) and in graptolites it consists of a SCLEROPROTEIN (i.e. a fibrous insoluble protein). Other invertebrates possess hard parts composed of a rigid intergrowth of crystals of CALCIUM CARBONATE in an organic matrix. The calcium carbonate may be ARAGONITE as in some molluscs, or CALCITE as in echinoderms. The skeleton in some of the simpler animals, like radiolarians and some sponges, is made of opaline SILICA occurring as discrete parts, SPICULES, or as a coherent meshwork.

The main structural elements in PLANTS consist of CELLULOSE, a polysaccharide with fibrous structure, and of LIGNIN, a complex aromatic substance.

Modes of preservation of fossils

It is exceptional for prehistoric animals to be preserved complete as were the refrigerated mammoths found in Siberia. Usually the soft tissues decay and the hard parts undergo changes which may affect the entire skeleton or only their organic components. Broadly speaking, the older the fossil, the greater is the alteration in the remains.

Changes in the organic component of hard parts

Fossils showing little change in the composition of the hard parts are likely to be of Cainozoic age. In these the organic components have decayed leaving the mineral substance of the shell, or bone, unaltered. Such fossils are porous and rather fragile, as for instance the mollusc shells from the Crag deposits in East Anglia.

Less commonly, fossils are found in which some of the organic matter remains, including rare traces of colour patterns (Plate I). Such fossils occur, as a rule, in deposits which are geologically unusual.

INSECTS IN AMBER. Many museums display pieces of amber from Oligocene deposits on the South Baltic coast which contain small fossils, mainly insects (Plate II). The amber is fossilised resin which entombed the insects as it oozed from coniferous trees. The chitinous skeletons are little altered, but the soft inner tissues are missing.

MUMMIFICATION. A small number of mummified animals is known in which both soft tissues and skeleton are preserved as a result of dehydration in a hot, dry climate. An example is the case of an extinct type of sloth which was found in New Mexico.

BONES IN PEAT OR TAR. Peat and tar have antiseptic properties which arrest the process of decay. The bones of Pleistocene animals entombed in these deposits are almost unchanged. Examples include bones of the giant deer from Irish peat bogs and of a vast array of extinct mammals from tar-pools in the United States.

CARBONISATION OF ORGANIC COMPONENTS. An organic substance like scleroprotein may be quite resistant to decay if it is sealed rapidly in sediment, and little-altered graptolite skeletons have been dissolved out of limestone (Plate III). More generally, organic components like scleroprotein, chitin, and the cellulose and lignin content of plants are CARBONISED, i.e. the relative carbon content is increased by liberation of volatile constituents.

The outline and, sometimes, details of the soft anatomy of an organism may be preserved as a carbonised residue as in the outlines of the body and tail of occasional fossils of ichthyosaurs (fig. 173).

Changes in the mineral substance of the hard parts

CONVERSION OF ARAGONITE TO CALCITE. Certain shells consist of aragonite which is liable, in time, to be converted to the more stable form, calcite. The recrystallisation involves the destruction of the structure of the shell while not affecting its main shape. Aragonite shells are common in Cainozoic rocks, less common in Mesozoic and are hardly known in earlier rocks.

PETRIFACTION. Petrifaction is literally the process of turning into stone; it involves the IMPREGNATION or REPLACEMENT of the hard parts by minerals deposited from solution in waters percolating through permeable remains. Impregnation is the infilling of the interstices left in a bone or shell on the decay of the organic matrix by an inorganic substance. Replacement is the substitution of a different mineral for the original mineral matter of the shell or bone. The resulting fossil has the outward form of the original skeleton, but is heavier, may differ in chemical composition, and in some cases the finer details of the skeletal structure are destroyed.

PETRIFYING MINERALS. The commonest petrifying substances are calcite, silica and iron compounds. Most frequently petrifaction is due to the impregnation of calcareous shells by CALCITE, as in echinoderms where the meshwork of pores in each plate is filled in by calcite resulting in a solidly crystalline plate which, if broken, shows cleavage faces. SILICA may replace calcite, chitin or wood. Amorphous silica (opal), in particular, may preserve the original microstructures in almost perfect detail (fig. 193). Silicified fossils which are enclosed in limestone can be dissolved from their matrix by dilute acid and so retain surface details

a   b

2   External mould and replica.

a, external moulds of a goniatite, *Gastrioceras*, Upper Carboniferous. b, a replica made by pouring a latex solution over the mould; once set the replica was peeled off (× 1·5).

3   Internal moulds in Portland Limestone.

The moulds are of (a) gastropod, and (b) bivalve shells which consisted of aragonite. Aragonite shells are more readily dissolved out of porous rock by percolating water than are calcite shells ($\times 0.8$).

like delicate spines which would normally be broken during extraction of the fossil from its matrix. Compounds of iron include HAEMATITE, LIMONITE (both oxides), SIDERITE (carbonate), PYRITES and MARCASITE (both sulphides). For the most part they replace calcareous shells, and the fossils tend to be poorly preserved, apart from those composed of marcasite or pyrites. Both the latter, however, tend to oxidise and disintegrate once exposed to air.

Loss of entire hard parts

Acidified water percolating through a permeable rock will dissolve and remove calcareous shells. The space left, the EXTERNAL MOULD, bears the surface markings of the shell in reverse. A latex (rubber) solution poured into this mould will provide a replica of the original shell (fig. 2). Sometimes a shell may have been filled with matrix before it was dissolved; this forms an INTERNAL MOULD (fig. 3).

TRACES OF SOFT PARTS. Impressions of the soft body of an animal, for instance a jellyfish or a worm (fig. 142), are sometimes found. More unusual fossils are the moulds of the intestines of some animals which have formed as a result of sediment filling and hardening in the gut before its decomposition.

TRACES OF ANIMAL ACTIVITIES. Animals may leave indications of their existence even when the organisms themselves have completely disappeared. These indications include footprints (fig. 165), mollusc tracks, worm burrows, bivalve (fig. 22) and sponge borings, tooth marks on bones, gastroliths (stones from a reptile's stomach), coprolites (fossilised faeces) and stone implements.

3

Occurrence of fossils

Rock types in which fossils occur

Fossils are mainly found in sedimentary rocks such as limestones, mudstones and sandstones which accumulated in former seas, lakes, deltas or river flood plains. They are most abundant in rocks laid down in relatively shallow seas, and are often sparse or lacking in rocks of continental origin. On rare occasions they may be found in other types of rock, for instance in travertine, tufa or siliceous sinter deposited from mineral springs; or, most unusual of all, in igneous rocks, as in the case of the carbonised tree trunk (Macculloch's Tree) standing in a Tertiary basalt in Mull.

LIMESTONES are often highly fossiliferous, sometimes consisting almost entirely of shells (fig. 4). These may be whole and uncrushed but more commonly are in fragments, and may have been rolled about so that delicate structures, like spines on brachiopods, are removed. The fine-grained rocks like CLAYS and SHALES may also be very fossiliferous, with well-preserved fossils showing fine details. In some instances the impressions of soft tissues may be preserved, a notable example being the superbly preserved details of soft-bodied creatures found as imprints in shales of Middle Cambrian age (Burgess Shale) in British Columbia (fig. 142). Most frequently fossils in these fine-grained rocks, especially the more delicate shells, are crushed by compaction, and sometimes may be distorted by tectonic forces. Clays and shales frequently contain NODULES which may consist of limestone, ironstone, phosphate or flint. Often these nodules have formed around organic remains and the fossils they contain are usually well preserved and uncrushed (figs. 5, 121).

Coarser-grained rocks like SANDSTONES are not commonly very fossiliferous apart from certain bands of sandstone, usually of marine origin and with a calcareous cement, which may be locally rich in fossils. The fossils are usually uncrushed but, owing to the normal permeability of sandstone, may be moulds or impressions (p. 12).

4   A shelly limestone with well-preserved ammonites of the genus *Astero-ceras*, Marston Limestone, Lower Jurassic, Yeovil ($\times 1 \cdot 3$). In places the shell has flaked off, exposing internal moulds.

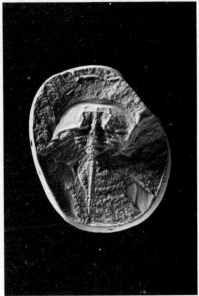

5 Fossils in ironstone nodules.

Left, *Euproops rotundata* (×0·8). Right, *Belinurus baldwini* (×1·6). Both xiphosurids
from the Upper Carboniferous, Rochdale. (Bronze replicas in the University of Nottingham.)

Physical conditions which determine the preservation of fossils

Observation of the various types of fossil-bearing rocks suggests that
fossilisation is favoured by particular conditions. The most obvious
requirements are an abundance of organisms, and some mechanism
which will ensure their rapid entombment on death. A good state of
preservation ensues if the remains are fossilised *in situ*, and therefore
are not subjected to abrasion by transport. Rapid burial prevents attack
on the dead organism by scavengers, abrasion by wave or current action,
and where the remains are covered by fine-grained sediment this may
seal them off and thus exclude oxygenated water which is necessary for
most processes of decay. Size, too, may be a factor of importance in
preservation. For instance, adult and therefore larger, forms may be
commoner as fossils owing to the more fragile nature of the skeleton in
young individuals. On the other hand, small size may sometimes be
advantageous. For instance, microscopic animals like foraminifera may
be overwhelmingly more abundant than larger. creatures. In a marine
environment their shells could be quickly buried (and thus finely
preserved) in great numbers while the large skeleton of a plesiosaur, for

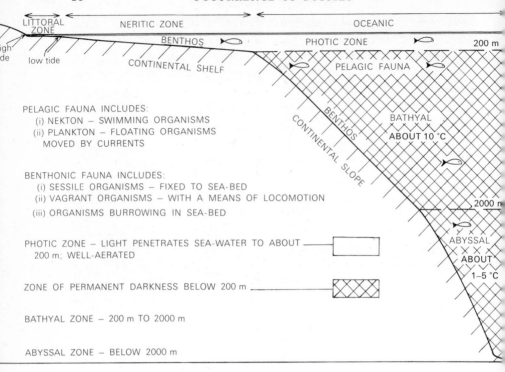

6   Distribution of major marine environments.

example, lying on the same sea-floor, would be exposed for a long time to all sorts of hazards before enough sediment accumulated to cover it.

Physical environment as a factor in fossilisation

Examination of the distribution of fossils in rocks shows that, in practice, the best conditions for preservation of organisms are met with in the sea (fig. 6), especially in coastal areas and on the continental shelf where organisms are abundant and sediments accumulate rapidly, and particularly in those parts of the sea where limestones form, e.g. in reefs. Here, preservation of shells is helped by the slightly alkaline environment which prevents their removal by solution. Conditions are less favourable in the littoral zone (fig. 6), however. Although organisms may be abundant here, their remains are subjected to attrition by wave action, and the constant shift of sediments by currents prevents rapid burial.

On land, accumulation of sediment is more intermittent than in the sea and rapid burial is fortuitous, depending perhaps on entombment during a major flood, or a sandstorm, in a marsh, a tar-pool or in resin (insects in amber). Usually the remains of an animal lie exposed to the elements, and to attack by scavengers, so that the skeleton distintegrates and, because of oxidising conditions, decay of animal and plant tissues is rapid. Thus terrestrial animals are rare as fossils and are seldom well preserved.

Transported fossils

The fossils occurring in a bed of rock may be of organisms which were preserved *in situ*. Such fossils are mainly of benthonic forms (fig. 6) and may be well-preserved and often entire specimens. In many cases, however, the fossils represent organisms from several different environments. A simple example is the association of pelagic forms (fig. 6) with a benthonic assemblage, or of land plants with marine shells. In the latter case, the plants must obviously have been transported, presumably by river agency. Further evidence of the transport of organisms is provided by broken, worn or sorted shells, by dissociated valves and

7   A derived fossil; an ammonite (*Mortoniceras*), from the Gault, Lower Cretaceous, of which internal moulds, in phosphate, occur in the Cambridge Greensand, Upper Cretaceous ($\times 2 \cdot 2$).

ossicles and by aligned shells. Shell fragments, like other mineral detritus, are sorted according to size or weight, so that a rock composed of equal-sized shell fragments is clear evidence of current action. Similarly, elongated shells subjected to current action are likely to be aligned.

Derived fossils

A bed of rock may contain, in addition to contemporary fossils, other fossils derived from an earlier formation (fig. 7). These derived fossils have survived the process of denudation which destroyed the rock in which they were originally enclosed; they are usually easily distinguished from the indigenous fossils by marked signs of abrasion due to transport, e.g. fragments of belemnite guards of Jurassic age occurring in Pleistocene gravels in south-east England.

Gaps in the fossil record

It has been pointed out that the most fossiliferous rocks are frequently those which accumulated in the relatively shallow parts of the sea. At the present day it is in this NERITIC zone (fig. 6), that life is richest, both in numbers and variety, and it is here that prompt burial is most likely to happen. Examination of the fossils occurring in rocks formed in this sort of environment shows that preservation of organisms has been selective, favouring forms which possessed hard parts – a skeleton or shell containing mineral matter. Animals consisting entirely of soft tissue are virtually unknown as fossils in these rocks despite their overwhelming abundance in life. Thus, worms burrowing in mud, sea-anemones attached to rocks, seaweeds and their attendant fauna, swimming creatures and plankton may vanish without trace, though their former existence must be suspected when examining the remains of a shallow-water fauna.

It is obvious that only a small fraction of the pageant of animal and plant history has actually been preserved and many factors have contributed to the incompleteness of the record. Living organisms may have been eaten by predators, dead organisms attacked by scavengers or their soft tissues decomposed by bacteria. Even hard skeletal parts are not indestructible, though complete decay may be a prolonged process. Preservation is finally ensured only if remains are rapidly sealed in sediment and left undisturbed while the processes of fossilisation operate.

Once formed, the later history of a fossil depends on what happens to its enclosing rock. It may be destroyed by metamorphism, by the process of weathering, or it may remain to be collected from an exposure by a geologist, perhaps by *you*. The fossils you collect, or examine in the laboratory are often very fragile and may break easily. In some cases, repeated handling can efface the surface markings by which the fossils are identified. Obviously, therefore, fossils should be examined with great care.

4

Fossils as indices of environment

Fossils may be invaluable in the reconstruction of the conditions in which sedimentary rocks formed. If the habitat of a living plant or animal is known it may be assumed that the same forms occurring as fossils also lived in similar conditions. Thus, for instance, since modern echinoderms live only in the sea, discovery of their remains in recent sedimentary rocks would clearly indicate that these had formed in the sea. Obviously deductions of this nature will be more certain in the case of the younger rocks which contain fossils closely related to modern forms. As we go back in geological time the relationship of fossil forms with present-day organisms becomes more remote, and extinct groups occur about whose environment nothing is directly known. In such cases, however, associated fossils which have living representatives may throw some light on the probable conditions in which the extinct forms lived. Here, however, the possibility must not be overlooked that a fossil may have been transported out of its habitat before being buried and fossilised. For instance, a land plant may drift down a river to the sea and eventually be buried in marine sediment along with marine organisms.

Most fossils are found in sediments which formed under water, and the range of aquatic environments is considerable, from fresh-water rivers and lakes or swamps, to saline lakes, brackish-water estuaries and lagoons, and the open sea. In each of these, conditions of salinity, temperature, light, depth, aeration and food supply may differ and will control the particular organisms to be found there. A sudden change, albeit temporary, in these conditions may cause mass mortality of the fauna, for instance the sudden lowering of temperature in a coastal lagoon during a spell of unusually cold weather. Quite small variations may be important in determining the presence or absence of a particular organism, and where these occur as fossils the appropriate conditions may be safely inferred.

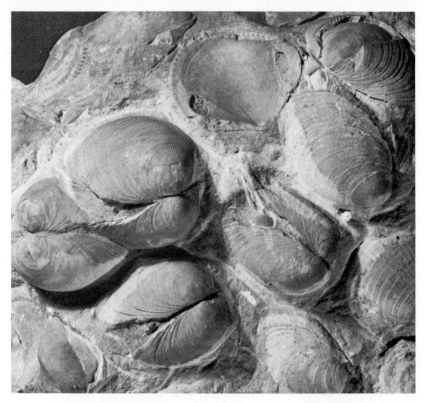

8   Fossils which indicate a shallow-water normal marine environment;
Glycymeris (bivalve), Eocene, Hampshire (×0·7).

In considering some of these factors in more detail, emphasis is placed
on aspects concerning marine environments.

Salinity

Salinity refers to the amount of dissolved salts, principally sodium
chloride, in water. Salinity is fairly constant in the open seas, about
$3\frac{1}{2}\%$; it is increased in closed seas where the rate of evaporation is high,
and is lowered in estuaries where river-water dilutes sea-water. Rivers
and most lakes are described as fresh-water.

Invertebrate groups which today are exclusively marine include
brachiopods, echinoderms, corals and most molluscs, the cephalopods
in particular; it can be assumed that their ancestors also lived in a marine
habitat (fig. 8). Extinct groups, like the trilobites, are constantly found
in association with known marine types, and so it is assumed that they

9   A fossil which indicates brackish water; *Potamides vagus* (gastropod), Middle Headon Beds, Oligocene, Isle of Wight ($\times 3$).

also were marine. Certain groups of fossils may be unreliable as indicators of normal salinity. Fish, for example, may live in the sea or in fresh water, or in both the sea and fresh water. Many Old Red Sandstone fish are regarded as fresh-water organisms because they are not found along with definitely marine creatures, and they occur in sediments quite different from the characteristic marine deposits of that time.

Some invertebrates are adapted to life in estuaries and lagoons where the salinity differs from that of normal sea-water. Two examples may be cited, the common mussel *Mytilus* (p.45) which frequently occurs attached to stones in muddy estuaries, and the gastropod *Potamides* (fig.9) which lives in estuaries and brackish-water swamps in tropical and subtropical areas; their occurrence in rocks without definitely marine fossils may be assumed to indicate brackish water. Sometimes, however, there is not enough evidence relating to salinity and in such circumstances fossils may simply be called 'non-marine'. This is the case with the two extinct forms shown in fig.10, which occur together in the

10   Non-marine fossils; right, ***Platyschisma helicites*** **(gastropod) and** ***Modiolopsis complanata*** **(bivalve), Lower Old Red Sandstone (×2).**

basal Downtonian, without any of the fully marine fossils such as are found in the underlying Silurian.

Living fresh-water invertebrates which have hard parts are restricted for the most part to a few genera of gastropods and bivalves. Similar types occur in the fossil record, especially in the newer rocks (fig.11), and may be associated with fish and plant remains.

Sunlight and depth of water

Sunlight is absorbed as it passes through water. In the sea only the upper region, the PHOTIC zone (fig.6), is illuminated. The depth of the photic zone varies according to latitude, season and clarity of the water. It is deeper in tropical areas where it may reach about 200 m (100 fathoms), but at this depth only the shorter wavelengths of light remain.

Plants depend on light for photosynthesis, and in the sea they are therefore restricted to the photic zone, most occurring at shallow depths.

Those plants which grow on the sea-floor are accordingly found in relatively shallow coastal water and their fossils give some general idea of the depth of the sea in which they grew, e.g. calcareous algae.

The restriction of plants to the photic zone determines the distribution of hordes of animals, including those which eat plants, e.g. herbivorous gastropods, the carnivores which eat the herbivores, the small fish and crustacea which shelter amongst the seaweeds, and the reef corals which have an intimate association with microscopic plants (p. 174).

Below the photic zone and extending down to about 2000 m is the BATHYAL environment. Various marine expeditions have shown the persistence in this completely dark region of a rich and varied fauna (fig. 12). It includes a benthonic fauna (fig. 6), mainly invertebrates, and a pelagic fauna including fish and squids. There is also a constant 'rain' of dead plankton (fig. 6) from the surface waters. This includes forms like diatoms and foraminifera, and is an important source of food for the organisms which live in the deeper regions. Records of bathyal deposits in the sedimentary column are of doubtful authenticity. How-

11   A fossil which indicates a fresh-water environment; *Viviparus* (gastropod), Eocene (×0·9). The inset is ×1·8.

I   A butterfly; *Prodryas persephone*, Oligocene, Colorado.

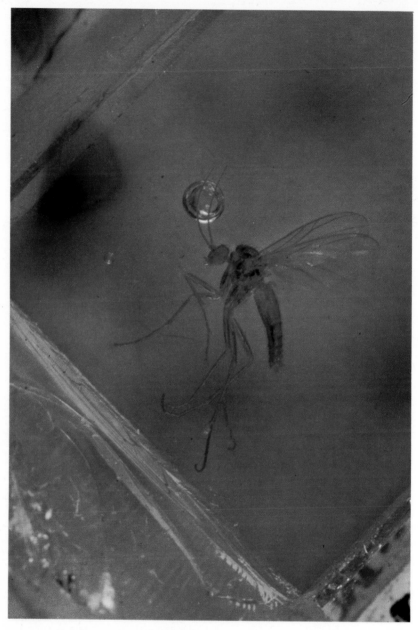

II   Fly in amber; *Eogonatium minutum*, Oligocene.

III. A graptolite; *Climacograptus*, Middle Ordovician.

IV A Triassic ammonite; *Monophyllites*, Hallstadt Limestone.

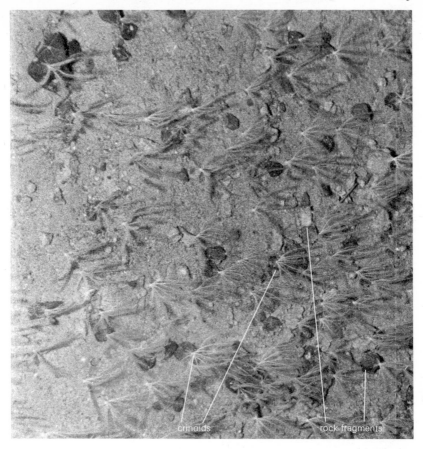

12   A modern deep-sea environment; a view of the ocean floor taken at bathyal depth, 651 m.

The bottom consists of shell sand and rock fragments, and the fauna includes many stemless crinoids. Locality: Galicia Bank, 42° 41′ N, 11° 35′ W; area shown, about 1 m square.

ever, certain sparsely fossiliferous rocks of Cretaceous and Tertiary age have been referred to this environment; their fossils are mainly of planktonic type.

Sea temperatures

The temperature at the surface of the sea at the present day ranges from about 30 °C in the tropics to 0 °C near the poles; it falls, too, with increasing depth and below the photic zone may be in the region of 10°–15 °C. The information about former sea temperatures which may be deduced from invertebrate fossils is necessarily fairly general.

13   Molluscs from the Pleistocene Crags of East Anglia.
a, *Scala greenlandica* (gastropod), and b, *Macoma calcarea* (bivalve) both of which live today in Arctic waters. c, d, *Acila cobboldiae* (bivalve, external and internal views, Lower Pleistocene only) which lives today in Japanese waters. (All ×1·8.)

Most types of invertebrates today tolerate a wide range of temperature, thus molluscs for instance may be found from tropical to arctic waters. However, particular species of molluscs are virtually restricted to either warm or cold water. The present distribution of certain of these molluscan species which also occur as fossils in the Plio–Pleistocene Crag deposits of Eastern England demonstrates the gradual deterioration of climate in Britain which culminated in the Ice Age. Broadly speaking, the percentage of those species now found only in northern cold waters increases towards the top of the Crag deposits, while the percentage of warm-water (Mediterranean) forms goes down (fig. 13).

Reef-forming corals are frequently cited as indicators of warm water, and indeed of a very precisely defined marine environment (fig. 106). Their preferred temperature range is 25°–29 °C, for reasons more fully explained later (p. 174). The reef corals of Tertiary and Mesozoic age are generally related to modern reef corals and may by analogy be used to indicate where subtropical climates occurred during that period of geological time.

Recent work has shown that the shell of some marine invertebrates may contain a built-in record of the temperature prevailing during the life of the organism. The shells in question are made of calcite and during growth the oxygen contained in the calcium carbonate includes two isotopes, O^{16} and O^{18} in a ratio dependent on the temperature at the time of shell formation; there is more O^{18} in shells from warmer water. The most useful fossils for this purpose are those from the newer rocks which have not been materially altered by recrystallisation or secondary deposits; they include belemnite guards, bivalves like *Inoceramus* and oysters and some Pleistocene foraminifera.

The shells in some marine species, which consist mainly of calcite, may also contain small quantities of magnesium carbonate, the amount of this substance being greater in forms which live in warm water compared with those from cold waters.

Aeration

Oxygen is essential to the metabolism of the vast majority of organisms. In many parts of the sea, enough oxygen is present to satisfy even a dense population of animals. As oxygen is used up it is constantly replaced. During the day oxygen is evolved as a byproduct of the photosynthesis of plants and at all times it is absorbed from the air at the surface of the water, a process which is augmented by turbulence at the surface, e.g. by waves breaking. Thus, there is more oxygen near shores, coral reefs and the surface of the sea; lower down the oxygen content is depleted by the respiration of animals, and by the bacterial decay of organic matter which involves oxidation. So long as there is free circulation of water from the surface towards the sea-floor enough oxygen is available to sustain life. But if the circulation is very slow as in the case of basins with limited inflow, oxygen in the deeper water is used up more quickly than it is replenished and in extreme cases is entirely removed with consequent ANAEROBIC CONDITIONS.

Anaerobic conditions

Most organisms cannot exist in an anaerobic environment. ANAEROBIC

BACTERIA, however, do not need free oxygen because their energy may, for instance, be based on the reduction of dissolved sulphates to form sulphides, or on the breakdown of proteins in organic matter. Hydrogen sulphide is released in these processes, and this may reduce iron compounds in sediment to ferrous sulphide, which in due course is changed into pyrite or marcasite.

A modern occurrence of anaerobic bacteria is in the black mud which is accumulating on the sea-floor in parts of the Black Sea where there is virtually no circulation of the water. The mud owes its dark colour to finely divided ferrous sulphide and organic matter. The organic matter consists of the remains of plankton which have settled from the aerated surface water. It has been decomposed by ordinary bacteria until all the available oxygen has been exhausted, and at this point anaerobic bacteria continue the process of decay. A strong smell of hydrogen sulphide is characteristic of this phase.

An anaerobic environment of this nature is devoid of a bottom-living fauna including scavengers, and conditions may be favourable for the preservation of skeletons of pelagic animals. Instances of dark shales which are believed to have accumulated in poorly oxygenated or even anaerobic conditions include the graptolitic shales of the Lower Palaeozoic and goniatite shales of the Carboniferous. In both cases the fossils may sometimes be pyritised.

Climate

Climate is a decisive factor in the distribution of most living plants and animals. The information about ancient climates which may be deduced from fossils, however, is to a large extent circumstantial and cumulative. In respect of the more recent fossils, we may use the present as a key to the past with reasonable confidence, but evidence from the older fossils must be treated with caution.

PLANTS AS A KEY TO ANCIENT CLIMATE. Quite detailed information about the climate of the British Isles in the interglacial and postglacial periods has been obtained from the analysis of pollen grains from successive levels in peat or lake clays. The pollen grains belong to species of trees which still grow in Europe in conditions of quite narrow climatic range. Their sequence records the fluctuations of climate which occurred. The range is from cold and dry, when an arctic flora (including dwarf birch, fig. 14c) colonised the area from which the ice had retreated, to milder and wetter weather when a mixed deciduous forest became established (e.g. oak, hazel, fig. 14a).

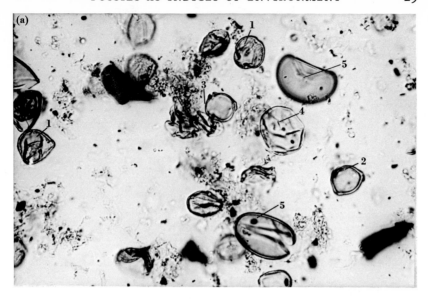

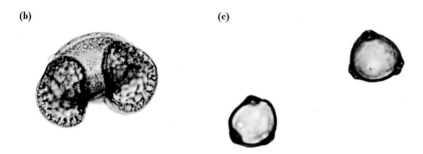

14   Pollen grains from Quaternary peats.

a, an assemblage of pollen grains typical of a temperate climate (1, oak; 2, hazel; 3, birch; 4, grass; 5, fern spores; ×270). b, pine pollen which indicates boreal conditions (i.e. warm dry summers, cold winters; ×120). c, birch pollen which indicates a colder climate.

Further back in time more reliance must be placed on modifications in plant structure which are related to humidity or to temperature and humidity. For instance, in some plants adapted to a dry habitat the leaves have a tough leathery texture with a heavy cuticle and sunken stomata (p.304), features which help to conserve water. Woody trees living in a region with a seasonal range in temperature and humidity show growth rings which are formed as a result of alternating periods of fast and slow growth. By contrast, in an area where the climate is equable, both warm and humid, trees grow steadily and growth rings are not developed. The absence of growth rings in *Cordaites*, a primitive woody tree found in the Coal Measures, accords with other evidence of the warm humid nature of the Coal Measure climate. It may be noted here that most of the plants in the Coal Measures reproduced by spores which required continued humidity for germination.

VERTEBRATES AS A KEY TO ANCIENT CLIMATE. The discovery of frozen woolly mammoths was of course a most impressive and indubitable proof of climate, in this case a period of prolonged glaciation. More indirect evidence of climatic conditions comes from fossil horses in late Miocene and Pliocene rocks. Their tooth structure shows that like the modern horse they grazed on grass. The distribution of their remains suggests that wide stretches of grassland existed in North America and Eurasia at that time; the main grasslands today are found in regions with a temperate climate of low rainfall.

In modern reptiles the body temperature is roughly that of their surroundings; they are restricted mainly to warm areas and are absent from arctic regions. This was probably also true of most reptiles during the Mesozoic when they were widely distributed throughout the world including regions (like the British Isles) which are now relatively cold. That the climate during this time was humid as well as warm is suggested by the abundance, and often great size, of plant-eating reptiles, which implies plentiful vegetation. Examples include the duck-billed dinosaurs, which were adapted to life in and around lakes and swamps.

Fossil fish might seem an unlikely source of information on climatic conditions. However, one group of fish, the lungfish (p. 250) are air-breathing forms which live in regions liable to seasonal drought. During the dry season they survive by burrowing in mud in the river bed. Remains of lungfish found in mud-filled vertical burrows in Permian rocks suggest that fossil lungfish too may have lived in a similar habitat.

PART TWO

The biology of fossils and their history

5

Foreword to the study of the different groups of fossils

The various groups of animals with which the palaeontologist is concerned are listed on pp. 6–7 in conventional order, beginning with the relatively simple Protozoa. In practice, it is convenient to adopt a different order for teaching purposes, depending, perhaps, on the sorts of fossils available locally, or simply on one's interests. The sequence followed in the succeeding chapters has a due regard for the maxim that what is familiar may be the more easily comprehended. Thus, the first groups dealt with are common at the present day, and the fossil examples described do not differ greatly from their living relatives. An added advantage is that fossils in this category are common in Mesozoic and Cainozoic rocks, where they are relatively easy to collect, and often are better preserved than fossils of greater age. The invertebrates, i.e. all animals other than the vertebrates, are by far the commonest as fossils and are of most importance stratigraphically. Accordingly, they are described first and in greater detail than the vertebrates and plants; forms of microscopic size (microfossils), however, are mentioned only briefly. The bivalves, the first group to be considered, are, together with the gastropods, the most important macroscopic invertebrates in the Cainozoic rocks; they are, also, the most familiar of seashore creatures today. The morphology of their shell is simple and functional, and they provide an ideal example of how past life may be interpreted with reference to present forms. It is logical to deal next with the other molluscs, the gastropods and the cephalopods, the latter containing extinct forms whose interpretation involves an element of conjecture. This aspect of palaeontology is also involved in dealing with the echinoderms, brachiopods and coelenterates, each of which has members confined to the Palaeozoic. The student is now well prepared for an examination of the extinct trilobites and graptolites, which involves a larger amount of conjecture.

The sequence of invertebrates just outlined is, of course, merely a suggested order, and readers need not be deterred from following their own preferences.

Choice of examples

The majority of fossils described here are found commonly in British rocks, and diagrams or photographs of most are included.

Classification

No attempt has been made to provide a detailed classification. At this level of treatment, it is enough to recognise the main biological group to which a fossil belongs, e.g. that *Cardium* belongs to the class Bivalvia. It is convenient, however, to have a term which embraces a number of closely allied genera, while, at the same time, clearly relating them to a genus already described. For this purpose, the name of that genus, but with the termination 'id' is used (or in some instances, 'oid'). Thus, the term 'agnostid' is used to denote forms closely related to the trilobite genus *Agnostus*, and similarly the term 'cidaroid' embraces the close relatives of the echinoid *Cidaris*. In a similar way, a generic name may be used in a broad sense as opposed to the strict definition. This is indicated by enclosing the name in inverted commas, e.g. '*Trigonia*' or by adding the letters *s.l.* after the name, meaning *sensu lato*; the strict definition is indicated by adding the letters *s.s.* meaning *sensu stricto*. It should be noted that the stratigraphical range of the genus *sensu lato* will be greater than that of the genus *sensu stricto*.

6

Mollusca, class Bivalvia

The Mollusca are soft-bodied invertebrate creatures most of which secrete a hard external shell. They include many familiar living forms like the whelks and limpets, cockles and mussels, the octopus and the cuttlefish, but many are known only as fossils.

Three CLASSES of molluscs are important as fossils, the Bivalvia (sometimes called Lamellibranchia), the Gastropoda and the Cephalopoda. Most are mobile creatures, living in the sea, but some are found in fresh water; only the Gastropoda occur on land. While externally the molluscan shell is very varied in form, internally the body largely conforms to a basic plan of organisation. In general they are bilaterally symmetrical except for the gastropods in which the body is twisted. There are four main regions, the head, the body, the mantle and the foot. The body (visceral mass) containing the internal organs (heart, liver, etc.) is DORSAL (uppermost) in position; on the VENTRAL (lower) side is usually a muscular organ used for moving, and hence known as the foot, except in the cephalopods in which it is modified. A sheet of tissue, the MANTLE, is attached to the visceral mass and hangs freely from it like a cloak; its outer surface secretes the shell which is partly calcareous and partly organic in composition. Within the cavity enclosed by the mantle are the gills. The mouth is at the ANTERIOR end; it leads into the digestive tract which in turn leads to the anus at the POSTERIOR end. The nervous and circulatory systems are highly organised throughout the phylum.

The marine molluscs have a world-wide distribution. A few are found in the deepest parts of the ocean, but the vast majority live on the seabed in relatively shallow water, and a similar distribution seems to have been true of most fossil molluscs. The phylum is abundantly represented in the fossil record from Cambrian times onwards.

The molluscs range in size from almost microscopic to gigantic; the giant squid, the largest known invertebrate, spans over 15 m.

CLASS BIVALVIA

(Alternative names: Lamellibranchia, Pelecypoda)

The bivalves are distinguished from other molluscs by their laterally compressed body enclosed between two calcareous valves which are united on the dorsal side by an elastic horny ligament. In most cases the animal is bilaterally symmetrical and the plane of symmetry is the plane along which the two valves meet. Thus the valves are almost perfect mirror images of each other. The function of the shell is to protect the soft edible body from some at least of the numerous carnivores of the sea. The group is entirely aquatic (marine and fresh water), the vast majority living a relatively sedentary life in shallow waters, never moving far or quickly, and following one of a limited number of modes of life. Modern examples of bivalves include cockles, mussels and oysters.

Morphology

Usually the bivalve shell completely encloses the soft body. The shell, consisting of two valves united by an elastic ligament, is secreted by the mantle; this is a fleshy tissue which hangs down as two folds, one on the right side of the body, and one on the left (fig. 15b). The valves are described as RIGHT and LEFT. They are united on the DORSAL surface of the body by the LIGAMENT and they separate along the other margins; these are distinguished as the ANTERIOR (where the mouth is

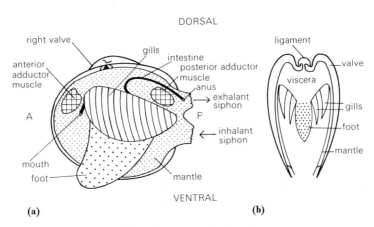

15   Morphology of the bivalves.
a, simplified diagram of a bivalve shell with the left valve and the left flap of the mantle removed. A, anterior; P, posterior. b, simplified section through a shell transverse to the plane of symmetry.

situated), the VENTRAL (opposite the dorsal), and the POSTERIOR (where the anus opens).

SOFT PARTS. The main mass of the body, including the viscera and organs such as the heart, lies in the dorsal part of the shell (fig. 15a, b). The foot, lying between the mantle lobes on the ventral side of the body, is a muscular organ which can be extended outside the shell and by alternately lengthening and contracting can pull the shell through soft sediment. Also within the mantle cavity lie the gills. Typically these function not merely as a respiratory organ, but also as a mechanism to collect food. On the surface of the gills are tiny cilia (filaments) which by beating backwards and forwards cause a current of water to pass through the mantle cavity. Incoming water is separated from the outgoing water because the mantle margins, which may be fused locally, open posteriorly at two points to form two SIPHONS. The current of water enters via the lower, the INHALANT siphon and particles of food (microscopic plants and animals) are sieved out by the gills and passed forwards in strings of mucus to the mouth at the anterior end. The outgoing current with waste products is passed out through the EXHALANT siphon.

SHELL. The substance of the shell is partly calcareous, partly organic (conchiolin, a proteinaceous material); a thin section shows it to consist of three layers (fig. 18e). On the outside is a thin horny layer, the PERIOSTRACUM. The inner layers consist of crystals of either calcite or aragonite, or both. Their structure varies in different bivalves. A common structure found in the middle layer consists of prisms lying transverse to the periostracum. The innermost layer in many shells is made of NACRE, otherwise known as mother-of-pearl. Nacre is made of alternate, very thin lamellae of aragonite and conchiolin lying parallel to the outer layers. A structure seen in many other shells consists of lath-shaped lamellae arranged in a regular alternating series; these shells have a dull porcellanous appearance. The outer two layers are secreted by the edge of the mantle; the innermost layer is laid down by the whole surface of the mantle and continues to grow during the life of the animal so that it may be quite thick compared with the outer two layers (fig. 16).

The soft parts affect the form of the shell to a minor extent only and while the shell itself may vary in shape it shows few structures. The apex of each valve, the UMBO, is on the dorsal side and typically lies in front of the ligament (fig. 18a). The umbones represent the earliest parts of the shell and, as the animal grows, additional increments of shell are laid down by the mantle and these are visible on the outer surface as concentric lines of growth (fig. 18b). While the surface of most bivalve

body cavity

a

b

16   Section of a bivalve shell.

a, the outer, relatively thin layer of shell secreted during growth by the edge of the mantle;
b, the inner, thick layer of shell laid down as a series of laminae by the whole surface of the
mantle. *Gryphaea*, Lower Jurassic (×2).

shells is relatively smooth apart from the fine lines of growth, in many
there is an ORNAMENT consisting of radial or concentric markings or a
combination of these. Concentric markings vary from fine growth lines
to quite coarse lamellae of regular or sometimes quite irregular appearance. Radial ornament varies from fine lines to coarse ribs and grooves.
Sometimes the radial and concentric elements are combined to produce
a reticulate pattern; occasionally spines or tubercles are present (fig. 17).

On the INNER surface of the valves is the HINGE PLATE which is a
thickening of the dorsal margin just below the umbo (fig. 18a). On each
plate are projections, TEETH, which fit into SOCKETS in the opposite
hinge plate. The teeth under the umbo are the CARDINAL teeth and
those beyond it are the LATERAL teeth. The teeth and sockets together

17 A spinose bivalve. Spondylus, Cretaceous (× 1·5).

make up the DENTITION and form a simple mechanism which guides
the valves to their correct positions relative to one another as they open
and shut. Some of the commoner types of dentition are defined on
page 57.

The remaining feature of the hinge apparatus is the LIGAMENT which
consists of resilient conchiolin. It is dorsal in position and it may be
EXTERNAL, lying above the hinge plate, or INTERNAL lying between
the hinge plates. In general, an external ligament lies behind the umbo,
a condition described as OPISTHODETIC (fig. 18a); in some forms it
extends in front of as well as behind the umbo, and this condition is

described as AMPHIDETIC (fig. 28f). In those forms with an internal ligament it may be situated in one pit or in a series of pits along the hinge plate (fig. 26h).

The hinge plate, with its dentition and ligament, is only a part of the mechanism by which the valves are opened and shut. The resilient ligament by itself would keep the valves gaping; they are closed by the contraction of ADDUCTOR muscles (fig. 18c). When these muscles relax, the external ligament, which has been under tension, pulls the valves apart (fig. 18d). In forms with an internal ligament, the ligament is under compression and thus pushes the valves open when the muscles relax.

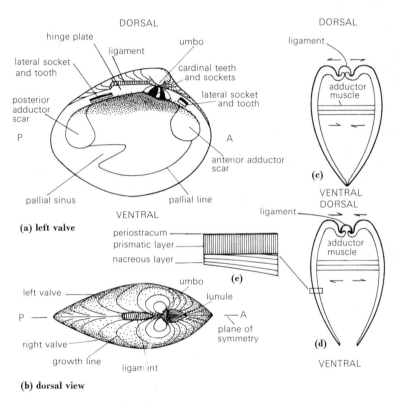

18 Morphology of the bivalves.

a, interior view of the left valve of an equivalve and inequilateral shell; in this and later diagrams which show dentition, the sockets are black and the teeth unshaded. b, dorsal view of the shell. c, section through a closed shell showing the adductor muscles contracted and the ligament stretched. d, through an open shell with the adductor muscles and ligament relaxed. e, diagrammatic section of a shell fragment, much enlarged, to show a type of microscopic shell structure commonly found in bivalves.

The points where the adductor muscles are attached to the valves are marked by SCARS which can be clearly seen on the inner surface of each valve; typically there is one at the anterior end (ANTERIOR ADDUCTOR), and one at the posterior end (POSTERIOR ADDUCTOR) of each valve. Shells with scars of roughly equal size are described as ISOMYARIAN (fig. 18a) while those in which the anterior one is smaller are ANISO-MYARIAN (fig. 23d); shells lacking the anterior scar are MONOMYARIAN and in these the posterior scar is enlarged (fig. 23c, f). A thin groove runs parallel to the ventral margin from the anterior muscle scar to the posterior muscle scar: this is the PALLIAL LINE along which the mantle is attached to the valves (pallium is another name for mantle) (fig. 18a). Bivalves which burrow in sediment have elongated tubular siphons which are extended from the burrow during feeding, and can be

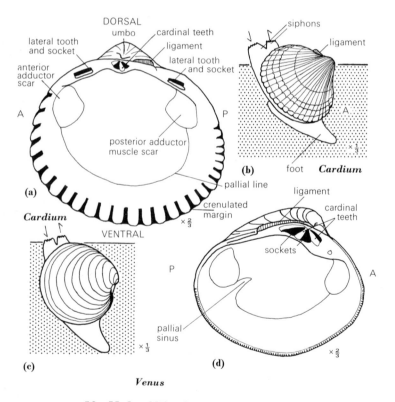

19 Mode of life of two common bivalves.

a, b, *Cardium*, a vagrant bivalve; a, interior of right valve; b, animal in feeding position in sandy sediment. c, d, *Venus*, a shallow-burrowing bivalve; c, animal in feeding position; d, interior of left valve. Arrows indicate the direction of the incurrent and excurrent flow of water.

retracted into the shell quickly if the animal is disturbed. The pallial line in such forms is not entire but shows an embayment, the PALLIAL SINUS, at the posterior end (fig. 18a). The depth of the sinus gives some indication of the length of the siphons and thus of the depth of burrowing.

When the valves are closed, typically the margins are pressed tightly together. Bivalves which live in burrows, however, may have a permanent opening, a GAPE, at the posterior end for the siphons (fig. 31c), and there may be a similar gape at the anterior end for the foot. The valve margins are usually quite smooth, but accurate closing is aided in some forms by the development of small crenulations of the margin (fig. 19a).

Most bivalves are bilaterally symmetrical about a plane passing between the two valves and these shells are said to be EQUIVALVE (fig. 18b). Each valve is, however, usually asymmetrical about a line from the umbo to the ventral margin and is said to be INEQUILATERAL (fig. 18a); in the majority of bivalves the umbo is nearer to the anterior end than to the posterior end. Thus the right and left valves can be readily distinguished; if the shell is held dorsal surface up and the anterior end pointing away from you, the right valve is then on your right and the left valve on your left. There are, however, exceptions to this rule, e.g. *Nucula*.

Shell form and mode of life of some common bivalves

Bivalves are adapted to a wide range of habitat. This has brought about modifications of the basic anatomy which is reflected in the form of shell. Shell shape may, therefore, be a clue to the mode of life of the bivalve and the forms described in the following pages illustrate some of the more usual modes of life. Their modifications can be paralleled among fossil bivalves.

Marine bivalves

Cardium (fig. 19a, b). The shell is rounded in outline, equivalve, and slightly inequilateral; rather globular in cross-section; the surface is strongly ornamented with radial ribs. The umbones are prominent, and curve inwards; the dentition is heterodont (p. 57); the ligament is opisthodetic; the muscle scars are equal in size; the pallial line is entire; the ventral margin is crenulate.

Miocene to Recent.

Many bivalves lead a vagrant life, moving sluggishly at or near the

surface of soft sediment. *Cardium edule*, the cockle, is an example found
on the lower shore of many sandy beaches where, for much of the time,
it lies buried to the depth of its shell and, when the tide is in, feeds by
extending its short siphons into the water (fig. 19b). When moving,
the cockle opens the shell slightly and extends the foot as a sharp-
pointed probe which elongates as it is thrust into the sand. It then
dilates so that the tip is anchored and its powerful muscles contract to
drag the cumbersome shell slowly through the sand. Unlike most vagrant

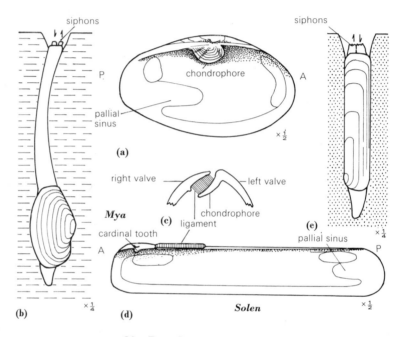

20   Deep-burrowing bivalves.

a–c, *Mya;* a, interior of left valve; b, the animal in its burrow in muddy sand with the siphons
extended; c, transverse section through the ligament and chondrophores. d, e, *Solen;* d, interior
of right valve; e, the animal in its feeding position at the top of its burrow in sand.

bivalves, the cockle can also move along the surface of the sand in a
series of hops; first the foot is bent almost double under the shell, then
with the tip pressed against the sand it is straightened suddenly jerking
the animal forwards.

Venus (fig. 19c, d). The shell is roughly oval in outline and is more compressed than that of *Cardium*; the surface ornament consists of concentric lamellae and faint radial lines. The umbones point forwards; the
dentition is heterodont (though the lateral teeth are minute); the liga-

ment is opisthodetic; the muscle scars are equal in size; a shallow pallial sinus is present; the ventral margin is crenulate.

Oligocene to Recent.

Venus burrows actively at shallow depths in soft sandy sediment. The shell, more compressed than the cockle shell, is moved more easily and quickly through the sediment. When feeding, *Venus* lies near the surface (fig. 19c), but when the tide goes out it burrows more deeply.

Solen (fig. 20d, e). The shell is thin and is elongated parallel to the hinge line (it may be eight times as long as broad); the umbones are near the anterior end; there are faint growth lines; there is a gape at both anterior and posterior ends. There are small cardinal teeth; the ligament is opisthodetic; the shell is isomyarian; the pallial sinus is short.

Eocene to Recent.

Solen is one of the 'razor' shells which burrow actively at some depth in sand. Since the shell gapes at each end the foot and siphons can protrude from the otherwise closed shell. While feeding *Solen* is at the top of its burrow (fig. 20e), but when the tide goes out, or it is disturbed, it digs downwards with its large and powerful foot. The smooth compressed shell facilitates very fast movement up or down through the sand.

Mya (fig. 20a–c). The shell is oblong, with almost central umbones and somewhat truncated posterior end with a wide gape; there are faint growth lines on the surface of the shell. There are no teeth; the internal ligament is supported by chondrophores (defined p. 57); the shell is isomyarian; the pallial sinus is deep.

Oligocene to Recent.

Mya lives in a deep burrow, 30 cm deep, in sand or muddy sand. It stays at the bottom of this burrow, and its siphons are long (they may be twice or three times the length of the shell) and are encased by a leathery sheath. They are not completely retracted and there is a permanent gape at the posterior end to accommodate them.

Pholas (fig. 21a). The shell is elongated and almost cylindrical; it gapes at both ends. The outer surface, especially the anterior end, is covered with rows of spines rather like a rasp. There are no teeth, and no ligament; the hinge line is in part reflected over the umbones to form a smooth rounded surface; there is a pallial sinus; an unusual feature is a short internal projection under the umbo to which muscles which operate the foot are attached.

Cretaceous to Recent.

Pholas, commonly known as the piddock, is found boring into a variety of rock types, including slate, sandstone, chalk, and peat (submerged forest) along the coasts of England. It bores by mechanical means. The foot is modified to form a 'sucker' which grips the inner end of its boring. The rounded hinge area provides a fulcrum on which the valves seesaw as the adductor muscles, each in turn, alternately contract and relax. This action rasps the spiny ornament of the shell against the rock, and as it rasps, the shell is rotated by a change in position of the foot. The result is a beautifully symmetrical cavity, wider at the inner end and constricted at the entrance. Internal moulds of the cavities made by fossil creatures similar to *Pholas* are known from British rocks of various ages (fig. 22), and may be associated with unconformities.

Teredo (fig. 21b). The shell is very much reduced and the animal with its naked and worm-like body is scarcely recognisable as a bivalve. A detailed description is unnecessary for it is of no importance as a fossil. The shell is used as a cutting tool to tunnel into drift-wood. Some of the cellulose of the wood is digested so that the wood is more than merely shelter. It is intriguing to find that even in Mesozoic seas, drifting timber was riddled with holes made by related 'shipworms'.

Eocene to Recent.

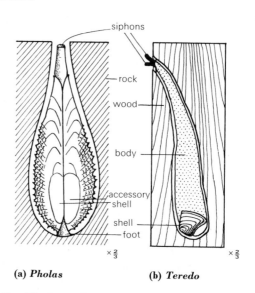

(a) *Pholas*   (b) *Teredo*

21   Bivalves which bore into hard material.
a, dorsal view of *Pholas* in its cavity in rock. b, *Teredo* in its tunnel in wood.

Mytilus (fig. 23d,e). The shell is very inequilateral; the umbones are at the sharply pointed anterior end; the posterior end is enlarged and rounded; there are faint growth lines. The dentition is poorly developed or absent; the ligament is opisthodetic; the muscle scars are unequal, the posterior one being the larger; the pallial line is entire.

Upper Jurassic to Recent.

Mytilus edulis, the common mussel, lives in the littoral zone on rocky shores and in estuaries. Here, although buffeted by rough waves and

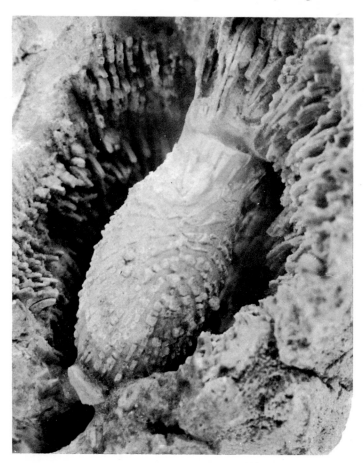

22 Internal mould of a cavity bored in coral by a bivalve, *Lithophaga*, Coral Rag, Jurassic, Upware (×4).

The cavity was drilled in the skeleton of a coral. After the death of the bivalve, the cavity and the interseptal spaces of the coral were infilled with calcareous sediment. Later, the coral skeleton (composed of aragonite) was dissolved away, leaving the space around the mould of the cavity.

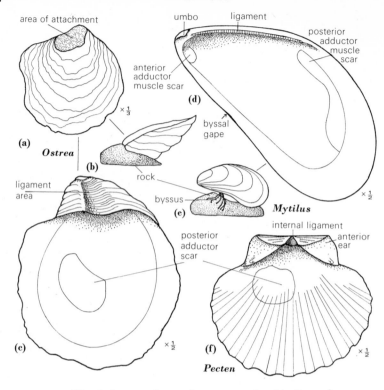

23　Anisomyarian and monomyarian bivalves.

a–c, *Ostrea;* a, external view of the left valve; b, a shell attached to rock; c, interior of the left valve. d, e, *Mytilus;* d, interior of the right valve; e, a shell attached to rock by its byssus. f, *Pecten;* interior of left valve.

swept by currents, it remains anchored to a rock by means of a *byssus*. This consists of horny threads (conchiolin) which are formed from a sticky fluid poured out from a gland in the foot on to a rock. The fluid hardens almost at once to form a thread and a succession of such threads is spun until the shell is secured to the rock.

Ostrea (fig. 23a–c). The shell is thick and irregularly circular in shape; it is inequivalve, the left valve being larger and convex while the right valve is more or less flat; the surface is roughly ornamented with irregular ribs and concentric lamellae. The hinge line is short and curved; there are no teeth; the ligament is in a triangular pit under the umbo; the shell is monomyarian, the posterior muscle scar being large and almost central in position.

　　Trias to Recent.

Ostrea lives with the left valve of its shell cemented to a hard surface (a stone or shell, see fig. 24) on the sea-floor. The common oyster, *Ostrea edulis*, is found in shallow water just below low tide level. When the young oyster emerges as a larva it swims around freely for a time. It then settles down on a clean surface to which the shell becomes fixed

24 Area of attachment of a fossil 'oyster' which settled on an ammonite shell. The area of attachment reflects the pattern of the ammonite in reverse. *Exogyra*, Cambridge Greensand, Upper Cretaceous (×1·7).

by means of a calcareous cement poured out from the modified byssus gland.

Pecten (fig. 23f). The shell is subcircular, almost equilateral, and it is inequivalve, the right valve being rounded and the left valve flat; the surface is ornamented by coarse radial ribs. The hinge line is short, straight and has 'ears' at either end; there are no teeth; the ligament is a triangular cushion sited in an internal pit on the hinge plate. The shell is monomyarian with a large, centrally placed muscle scar.

Eocene to Recent. Many allied genera range from the Trias.

Pecten lives a partly sedentary, partly free-swimming life. *Pecten maximus*, best known as the 'scallop' shell, is found in shallow water around Britain. When very small, *Pecten* is attached by a byssus. As it grows larger, *Pecten* discards the byssus and lies free on its right valve. *Pecten* can swim for short distances, albeit in a somewhat inelegant manner, by flapping its valves together. It alters the direction of its movement by manipulation of the mantle edges. The swimming action is made possible by the organisation of the hinge line and the adductor muscle. The stout ligament causes the valves to gape widely, about 30°, as the muscle relaxes; then, as the muscle contracts quickly to close the valves, any water in the mantle cavity is ejected and thrusts the shell in the opposite direction.

Fresh-water bivalve

Unio (fig. 26c) has an elongated oval shell with an inner layer of nacre. There are several more or less horizontal cardinal teeth; the ligament is external and opisthodetic. The muscle scars are deep and equal in size; the pallial line is entire.

Trias to Recent.

Unio is a fresh-water bivalve; it occurs in rivers where it ploughs slowly forwards through sediment. In water of low pH the umbones are often eroded. The larva is for a time parasitic on fish.

SUMMARY. The form of the bivalve shell is to a considerable extent a guide to the mode of life. Thus, normal vagrant bivalves, which move through the sediments at or near the surface of the sea-floor, have a shell which is equivalve and inequilateral; in addition the shell is isomyarian and has an entire pallial line. Bivalves which burrow in soft sediment are characterised by a pallial sinus and typically the shell is elongate and compressed; in many forms the shell gapes and the dentition may be poorly developed. Forms which bore into hard rock have a shell resembling that of the burrowing type except that it is more or less cylindrical. Bivalves which are attached by a byssus usually have a very inequilateral shell with the umbones at or near the anterior end, and the posterior end much enlarged; the shell is anisomyarian. Bivalves which are cemented to the sea-floor have an inequivalve shell with the fixed valve the larger, and the free valve flat and lid-like; the shell is generally monomyarian.

Additional common genera

Most of the forms described below are grouped according to the era in which they first appeared, but some have a more extended range, e.g. *Modiolus*, described under the heading 'Palaeozoic bivalves', is still common at the present day.

Palaeozoic bivalves

Ctenodonta (fig. 25b). The shell is oval, with faint concentric striae. The ligament is external; the dentition is taxodont; the pallial line is entire.
Ordovician.

'*Nucula*' resembles *Ctenodonta* but has an internal ligament. It differs from most bivalves in having the umbones nearer the posterior end. A related form is shown in fig. 13c, d.
Silurian to Recent. *Nucula s.s.*, Cretaceous to Recent.

Modiolus (fig. 25a). The shell is equivalve and very inequilateral with the hinge line long and straight; there are no teeth. The shell is anisomyarian.
Devonian to Recent.

Dunbarella (fig. 29a) is a thin-shelled form resembling *Pecten* in shape. The ornament consists of numerous fine radial and concentric lines. The hinge line is wide.
Carboniferous.

Carbonicola (fig. 29b) resembles the modern *Unio*. It has a thick hinge plate which may have no teeth or may have one or two cardinal teeth.
Lower Coal Measures.

Mesozoic bivalves

Trigonia (fig. 26a), has a thick shell with a nacreous interior. The shell is very inequilateral; it has a coarse ornament, varying according to species, of concentric ribs or tubercles. The dentition is schizodont, and

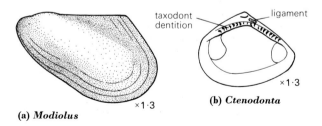

(a) *Modiolus* ×1·3

(b) *Ctenodonta* ×1·3

25 Palaeozoic bivalves.

the teeth are grooved. The pallial line is entire. '*Trigonia*' lives today in Australian waters.

Mesozoic. *Trigonia s.l.*, Mesozoic to Recent.

Gryphaea (figs. 16, 26g) belongs to the oyster family. The left valve is very much thickened and the umbo is strongly incurved over the hinge line; the right valve is more or less flat. *Gryphaea* was attached by its left umbo in early life; the adult was not cemented but lay free with its left valve resting in mud.

Trias to Eocene.

Exogyra (fig. 26d, e) also belongs to the oyster family. Its distinctive feature is the spiral twisting of the umbones in the plane of the hinge line.

Jurassic to Eocene.

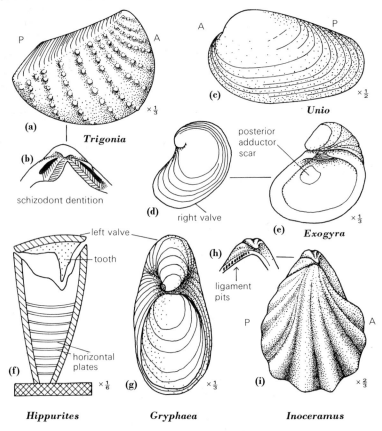

26   Mesozoic bivalves.

Pholadomya (fig. 31c) is a thin-shelled burrowing form with a wide gape at the posterior end. The ornament, of radiating ribs and concentric folds, fades out towards the posterior end. Living forms are found in warm water.

Trias to Recent.

Inoceramus (figs. 26i, 27a). The shell is inequivalve and inequilateral. The ornament varies in different species, the pattern being concentric in some and radial in others. There are no teeth; the hinge line is long and

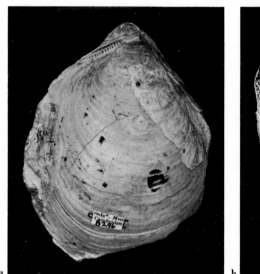

a   b

27   Mesozoic bivalves.

a, *Inoceramus*; the valves are slightly displaced, exposing the ligament pits along the hinge margin (×1·1). b, *Hippurites* (0·4). Both Upper Cretaceous.

has many transverse ligament pits; there may have been a byssus. The shell is much thickened in some later species.

Lias to Chalk.

Hippurites (figs. 26f, 27b) was fixed to the sea-floor by the apex of the right valve, and this valve is conical to cylindrical in shape and may be a foot or more in height; a series of horizontal plates cut off the lower part of the valve so that the animal occupied only a small space in the upper part. The free valve formed a lid which could be pushed up; peg-like teeth projected from it into sockets in the fixed valve. *Hippurites* is a late member of the Mesozoic rudists, aberrant sessile bivalves. The

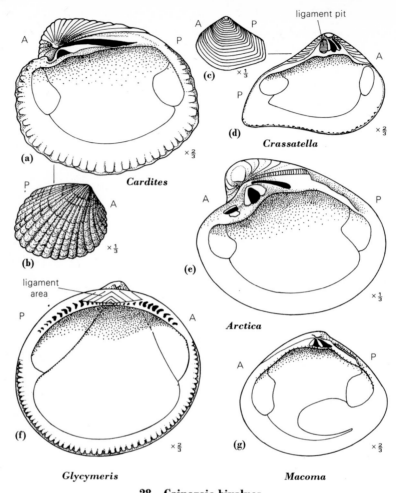

28 Cainozoic bivalves.
The dentition is heterodont in each case, except (f) where it is taxodont.

rudists occur mainly in the deposits of Tethys, a tropical or subtropical sea ancestral to the Mediterranean; they grew in reefs in the littoral zone. The later members like *Hippurites* show a superficial likeness to rugose corals, in the shape of the fixed valve with its horizontal plates.
Upper Cretaceous.

Cainozoic bivalves

Arctica (fig. 28e). The shell is roughly oval in shape, with prominent umbones. The dentition is heterodont; the pallial line is entire.
Cretaceous to Recent.

Cardites (fig.28a,b). This is a thick-shelled form with prominent umbones. There are broad radial ribs on the surface. The dentition is heterodont. The ventral margins are coarsely crenulated on the inside of the valves.

Eocene to Recent.

Glycymeris (fig.28f). The shell is almost equilateral. There may be radial striations on the surface. The ligament is amphidetic, and is inserted in grooves of inverted V-shape. The dentition is taxodont. The muscle scars are subequal and the pallial line is entire. The ventral margins are crenulated.

Cretaceous to Recent.

Crassatella (fig.28c,d). The shell is more or less oval in outline and is truncated at the posterior end. The surface is smooth or may have concentric furrows. The dentition is heterodont; there is a triangular ligament pit posterior to the teeth.

Cretaceous to Recent.

Geological history of the bivalves

SUMMARY. Bivalves are first recorded from rocks of Middle Cambrian age in Spain. Their fossils are uncommon until the Silurian, but between then and the end of the Palaeozoic many genera appeared, occurring abundantly in some horizons. A high percentage of forms died out at the close of the Permian and many new genera appeared in the Trias. A great diversity of forms appeared during the Mesozoic and the class reached its acme during the Tertiary; it remains a very important group.

Lower Palaeozoic bivalves are often poorly preserved, moulds being common. This may mean that the original shell was composed of aragonite rather than of calcite. (Aragonite is less stable than calcite.) Most of these early forms appear to have been vagrant forms which ploughed through soft sediment. *Ctenodonta* (fig.25b) and '*Nucula*' are typical examples; *Nucula* still survives (see *Acila*, fig.13c,d). Some forms appearing in the Silurian may have been burrowers and some were attached by byssus.

Upper Palaeozoic bivalves are on the whole better preserved, and the shell is usually present. Possibly the earliest fresh-water form occurs in the Old Red Sandstone. In the Lower Carboniferous, marine bivalves are common at some horizons. Vagrant forms are the most likely to be

a

b

c   d

29   Upper Carboniferous bivalves.

a, a marine form; *Dunbarella* (×1·5). b–d, non-marine forms; b, *Carbonicola*; c, *Naiadites*;
d, *Anthracosia* (all ×1·2).

30   A Cretaceous 'oyster', *Lopha*, Lower Chalk, Cherry Hinton (×1·7).
The characteristic feature of this form is the strong radial folding of the valve margins. Compared with other oysters, the zigzag suture between the valves must have made possible an increased flow of water, resulting in more efficient feeding and removal of waste.

found in the reef limestones and *Modiolus* (fig. 25a) is a byssally attached form occurring in limestone-shale 'lagoon' facies.

In the Millstone Grit series, impressions of *Pecten*-like shells in black shales are frequently associated with goniatites, e.g. *Dunbarella* (fig. 29a). In the Coal Measures bivalve shells may be thickly concentrated in 'mussel' bands. These forms are referred to as 'non-marine' because it is uncertain whether they lived in fresh or brackish water; some, e.g. *Carbonicola* (fig. 29b), were vagrant forms, and others were attached by byssus (fig. 29c). Many of these forms have a limited vertical distribution and a wide lateral dispersal in this facies so that they are useful as zonal fossils; a number of zones based on their occurrence have been recognised in the Coal Measures.

While a proportion of Palaeozoic genera survived into the Mesozoic,

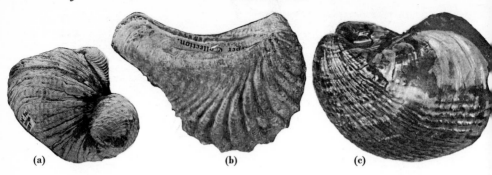

(a)   (b)   (c)

31 'Warm' water bivalves which are not found in England after the Cretaceous (b) or early Tertiary (a, c), but of which surviving species live in warm
regions.

a, *Chama*, Barton Beds, Eocene. b, '*Trigonia*', Upper Greensand, Blackdown. c, *Pholadomya*, London Clay (all ×0·9). (c, is tilted to show posterior gape.)

many more disappeared, and increasing numbers of new genera
appeared during the course of the Jurassic and Cretaceous. The preservation is generally good. So abundant are bivalves at some horizons that
whole or broken shells may make up most of the rock. The following
examples are mainly of unrelated genera which, at particular horizons,
found conditions favourable to their way of life, and thus, for a time,
were very prolific. *Trigonia* (fig. 26a) is a vagrant form common in some
horizons in the Jurassic; it survives today in warm waters around
Australia. *Ostrea* is the most frequently occurring cemented type, and
there are related forms common at several horizons; for instance
Gryphaea (fig. 26g), one of the few fossils to have a common name – the
devil's toenail – aptly suggesting the strongly convex and very much
thickened left valve. *Exogyra* (fig. 26d, e) and *Lopha* (fig. 30) are other
'oysters'; they are more typical of the Cretaceous. *Pholadomya* (fig. 31c),
often found in some of the Jurassic limestones is a burrowing form;
surviving species live in warm water. *Inoceramus* (fig. 26i) is an important
form in the Upper Chalk. The rudists (fig. 27b) are largely restricted to
deposits of the Tethyan region (p. 52).

Bivalves share with gastropods a predominant position in the shallow-
water faunas of the Tertiary rocks; there are many Tertiary forms still
living, the majority having heterodont dentition. Examples are shown in
fig. 28. In the earlier Tertiary rocks in England there are a number of
genera which today are restricted to warmer waters, e.g. *Chama* (fig. 31a),
Crassatella (fig. 28c, d). Towards the end of the Tertiary bivalves which

still live in British waters appeared in increasing numbers. These were accompanied in the Pleistocene by other forms which today are found only in colder northern waters, e.g. *Macoma calcarea,* a burrowing form (figs. 13b, 28g).

TECHNICAL TERMS

ADDUCTOR MUSCLE SCARS two scars, one anterior and one posterior, impressed in each valve by the muscles which close the valves. The scars are equal in size in isomyarian shells (fig. 18a); the posterior scar is larger than the anterior scar in anisomyarian shells (fig. 23d); there is only one scar, the posterior scar, in monomyarian shells (fig. 23c).

BYSSUS horny, fibrous outgrowth from the body by which the shell, in some forms, is secured to a firm surface (fig. 23e). A gape or notch may be present in the valve margin near the anterior end to accommodate the byssus when the valves are closed.

CHONDROPHORE a special device found in *Mya* (fig. 20a, c) to accommodate the internal ligament. In the left valve it is a spoon-shaped process projecting from the hinge line at about 90°; in the right valve it is a shallow pit under the umbo.

DENTITION the collective term for the teeth and sockets of the hinge plate. The principal types of dentition are:
(a) TAXODONT many small similar teeth and sockets all along the hinge plate (fig. 25b); (b) HETERODONT a few teeth varying in size and shape, distinguished as the CARDINAL teeth which radiate from the umbo, and the LATERAL teeth which lie obliquely along the hinge plate on both anterior and posterior sides of the umbo (fig. 19a); (c) SCHIZODONT two or three teeth, rather thick and sometimes grooved, lying under the umbo (fig. 26b).

EQUILATERAL the shell is more or less symmetrical about a line from the umbo to the ventral margin (fig. 28f). A shell in which the umbones are nearer the anterior end than the posterior end is inequilateral (fig. 18a).

EQUIVALVE the valves are mirror images of each other except for the minor differences on the hinge line due to the dentition (fig. 18). A shell in which the valves are dissimilar in size and shape is INEQUIVALVE (fig. 26g).

GAPE a permanent opening between the valve margins; a posterior gape occurs in some burrowing bivalves with a large siphon; an anterior gape through which the foot extends may also occur in some burrowers, e.g. *Mya* (fig. 20b).

HINGE PLATE a thickening of the dorsal margin, either straight or curved, along which the valves articulate. It supports the dentition and the ligament (fig. 18a).

LIGAMENT a resilient horny substance (formed from an organic material, conchiolin) which unites the two valves along the dorsal margin. It may lie posterior to the umbones, OPISTHODETIC (fig. 18a), or on both sides of the umbones, AMPHIDETIC (fig. 28f). It may be external or internal. An external ligament is under tension when the valves are shut and, being elastic, causes the valves to gape when the adductor muscles are relaxed. An internal ligament usually occurs in a pit (fig. 23f), or a series of pits (fig. 26h) in the hinge line. It is compressed when the valves are closed, and when the adductor muscles relax, it pushes the valves open (fig. 18c, d).

ORNAMENT surface markings on the outside of the shell which may be concentric about the umbo, or radial from the umbo, or both concentric and radial. The ornament may consist of striae, ribs, or occasionally of knobs or spines.

PALLIAL LINE a line running between the anterior and the posterior muscle scars, parallel to the ventral margin of the valve. It marks the attachment of the mantle to the valve and it may be entire (fig. 19a), or deflected by the PALLIAL SINUS (fig. 18a).

PALLIAL SINUS a bending inwards of the pallial line at the posterior end of the shell which is found in burrowers (fig. 18a).

UMBO (umbones) the earliest formed part of the valve, usually a rounded boss which projects above the hinge line (fig. 18a).

7

Mollusca, class Gastropoda

The distinguishing feature of the gastropods is the torsion of the viscera (internal organs). The body is asymmetric; there is a distinct head at the anterior end, and on the ventral surface is a muscular creeping foot. The body is protected in most forms by a single (univalve) shell (forms lacking a shell are not considered here). Typically the shell is a tapering tube, coiled in a right-handed spiral.

Gastropods are more abundant than any other group of molluscs at the present day; they also occupy a greater range of habitat. The majority are aquatic, and of these most live in shallow seas; they are also widespread in fresh water, and on dry land. Modern examples include the marine limpets, winkles and whelks, and the terrestrial snails and slugs.

Morphology

SOFT BODY. The shell is a refuge into which the entire body can be withdrawn. When the animal is moving, the head and foot are extended outside the shell but the soft visceral hump remains within the shell. The head bears sensory tentacles and eyes, and below these is the mouth in which is a food-rasping tool, the radula. This is a flexible horny ribbon bearing rows of teeth. The foot (fig. 32b) is an elongate organ with a flat sole on which the animal glides over the sea-floor by continuous waves of muscular contraction. Terrestrial forms lubricate the dry surface over which they move by exuding mucus.

The visceral hump lies on the dorsal surface of the body. It contains the digestive system and other organs. Much of it is spirally coiled within the shell. It is covered by the mantle which extends towards the head to form a space, the mantle cavity, which is thus anterior in position, in contrast to its normal position in other molluscs. Other parts of the body, including the anus, are also anterior in position because during the development of the larva the mantle and viscera are twisted through 180° relative to the head and foot.

In aquatic forms the mantle cavity generally contains feathery gills. These extract oxygen from water circulated through the cavity by ciliary action. In some forms the anterior margin of the mantle may form a tubular extension, the INHALANT SIPHON (fig. 32b) to direct water to the gills. Gills are lacking in terrestrial forms in which the mantle cavity is modified to act as a lung.

SHELL. An empty gastropod shell provides few clues to the internal structure of its former inhabitant. Most fossil shells can, however, be compared with fairly similar shells of living forms.

The shell consists mainly (96%) of crystals of calcium carbonate – usually aragonite, but sometimes calcite – intermeshed with organic

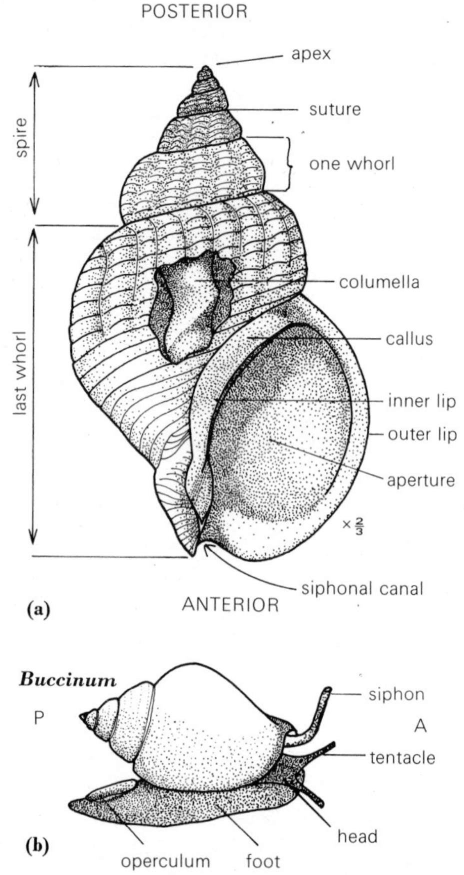

32 Morphology of the gastropods based on *Buccinum*.
a, shell with part of the last whorl broken to expose the columella. b, a gastropod in crawling position with the head and foot extended.

33   Longitudinal section of a gastropod showing the columella (×3).

matter; it is covered with a layer of periostracum. There are usually two layers in the shell, each with a distinct crystal arrangement which resembles the microstructures found in bivalve shells (*q.v.*, p.36), for instance nacreous and crossed lamellar structures.

The shell is basically a conical tube, closed at the pointed end, the apex, and open at the wide end, the APERTURE (fig. 32a). The form of the shell results from variations on a very few themes which, combined in different ways, produce distinctive shells for each of hundreds of species of gastropods. The themes include coiling, the rate of increase in diameter of the shell, the shape of the cross-section, form of aperture and ornament.

COILING. The shell, secreted by the mantle, grows by increments to the margins of the aperture; in most cases the increments are greater along one margin so that the shell coils about an axis. Coiling is typically in a spiral, like the thread of a screw, descending from apex to aperture; planospiral coiling is exceptional. In some gastropods coiling is confined to the embryonic shell (protoconch) at the apex, and the fully grown shell is a conical 'cap' shape. Coiling of the shell is, of course, independent of the torsion of the body which occurs in the larval stage (p. 59).

Each complete coil of a shell is a WHORL; the line along which successive whorls meet is the SUTURE. The ultimate whorl is the LAST WHORL, and the earlier whorls together form the SPIRE (fig. 32a). The shell may be coiled tightly about its axis so that a solid central pillar, the COLUMELLA, is formed. In loosely coiled shells there is an axial space which forms the UMBILICUS where it opens at the base of the last whorl.

The last whorl may be little larger than the previous whorl if the diameter of the shell increases slowly (fig. 35f); if the diameter of the shell increases rapidly the last whorl may be much larger than the spire (fig. 35d). The spire may be high, pointed and consist of many whorls (fig. 35f); it may be short, and have few whorls (fig. 35d); it may be depressed (fig. 37c) or occasionally it may be concealed by the last whorl (fig. 37a).

The aperture may be rounded, oval or slit-like. The margin nearest the apex is termed the POSTERIOR margin; the opposite side is the anterior margin, where the head emerges. The margin in contact with the previous whorl is the INNER lip; the free margin is the OUTER lip (fig. 32a).

The aperture may be entire (fig. 35d), or extended at the anterior margin by a SIPHONAL canal (fig. 38), or cut by an EXHALANT SLIT (fig. 37). The siphonal canal ranges from a short deflection of the anterior margin to a long narrow canal like a split tube. It supports the inhalant siphon (p. 60). The exhalant slit is a narrow slit in the outer lip at right angles to the edge. The earlier part is filled in by shell during growth; the resultant trace on the shell is called a SLIT BAND. An exhalant slit is found in living gastropods with two gills; it occurs in

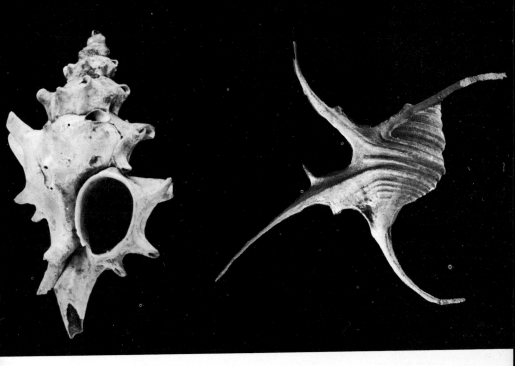

34 Gastropods with spinose processes.

Left, *Typhis*, Barton Beds, Eocene (×4); the spines are hollow tubes some of which are former siphonal canals. Right, '*Aporrhais*', Lower Cretaceous (×1·5); the spines are extensions of the outer lip which were resorbed during shell growth so that only those last formed remain.

many fossils, especially Palaeozoic forms. In some genera a subsequent layer of shell, CALLUS (fig. 35d), is deposited by the mantle on the inner lip and adjacent part of the whorl. In many marine gastropods a horny lid, the OPERCULUM (fig. 32b), closes the aperture when the body is withdrawn into the shell. It is borne on the posterior end of the foot; it is not preserved in fossils.

ORNAMENT. The surface of the shell may be smooth, or bear fine or coarse markings arranged transversely or spirally; knobs and spinose projections sometimes occur (fig. 34).

Internal markings are confined to scars left by the muscles which attach the animal to its shell. These are not usually seen except in cap-shaped shells (fig. 35c); in coiled shells they occur on the columella.

ORIENTATION. Most gastropod shells are asymmetric. The shell is conventionally drawn with the aperture facing you, and the apex of the spire pointing upwards. Most genera are coiled in a clockwise direction so that the aperture is on your right, DEXTRAL coiling (fig. 32a). Occasionally shells are coiled in a left-handed direction, sinistral coiling (fig. 38a), and the aperture is then on your left.

Shell form and mode of life of some common gastropods

Marine gastropods

Patella (fig. 35a–c). The shell is thick and is of widely conical shape with the apex nearer the anterior end. The aperture is oval. The ornament consists of ribs or striae radiating from the apex. There is a muscle scar of horseshoe shape on the inside of the shell, the open end being anterior.

Eocene to Recent.

The common limpet, *Patella vulgata*, lives mainly on rocky shores. At rest, *Patella* clings so tenaciously to its rock with the sole of its rounded foot that the roughest of waves do not dislodge it. Its low

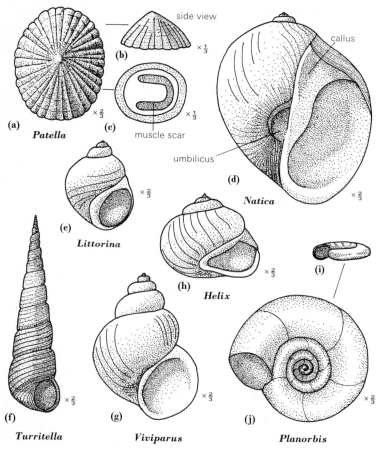

35   Gastropods with an entire aperture.

conical shape offers minimum resistance to waves, and when the tide is out its habit reduces desiccation. At high tide *Patella* crawls a short way in search of encrusting algae on which it browses and then returns 'home' to the oval-shaped depression it has worn in the rock.

Littorina (fig.35e) has a thick shell ranging in size up to about 2·5 cm. It is of rounded shape and has a low spire of a few whorls. The aperture is rounded, the outer lip is sharp. There is a columella. The surface is smooth or marked faintly by spiral lines.

Lias to Recent.

Littorina littorea is the commonest of the periwinkles found around the British coasts. It is a vagrant form occurring on both rocky and muddy shores. It may be out of water for many hours each day when the tide is out, but is in no danger of desiccation because the operculum closes the aperture so exactly. The thick shell is resistant to damage, however much it is rolled around in stormy weather. The periwinkles are herbivorous, feeding on seaweeds.

Buccinum (fig.32a) has a thick shell of roughly ovoid shape; the spire is moderately high and the last whorl large. There is no umbilicus. The aperture is oval; there is a short, bent siphonal canal. There is callus on the inner lip.

Pliocene to Recent

The common whelk, *Buccinum undatum*, ranges from the lower shore down to about 200 m. Inshore shells are about 5 cm length; in deep water larger shells up to 15 cm long occur. *Buccinum* ploughs through soft sediment with its inhalant siphon projecting above mud level to maintain a flow of clean water through the mantle cavity. It is carnivorous, rasping the soft flesh of its prey with its radula.

Natica (fig.35d) has a thick smooth shell of almost globular shape; the spire is short and the last whorl large. There is an umbilicus which may be partly obscured by callus. The aperture is entire.

Trias to Recent.

Natica alderi has a large foot (partly reflected over the shell), with which it moves through sand just below the surface, seeking small bivalves. *Natica* secretes acid to soften the shell of its prey, and bores a neat round hole through which the soft body is scooped out. The hole is small and tapers inwards; similar holes are often seen in fossil shells, e.g. in bivalves from the Red Crag (fig.36).

36   A hole drilled by a carnivorous gastropod, probably *Natica*, in a bivalve shell from the Red Crag, Pleistocene (×2).

Turritella (fig. 35f) has a turreted shell of many whorls which form a high pointed spire about 50 mm in length; it is ornamented by spiral ribbing. There is no umbilicus. The aperture is subquadrate and entire; this feature distinguishes *Turritella* from otherwise similar shells like *Cerithium* which have a siphonal canal.

Cretaceous to Recent.

Turritella communis is found in shallow water, about 22 m deep, where it burrows spire downwards in muddy gravel. Water is drawn into the mantle cavity by ciliary action; suspended food particles in it are collected by the gills, embedded in mucus and passed to the mouth.

Fresh-water gastropod

Planorbis (fig. 35j) has a thin shell which is almost planospiral; it is coiled so that a depression is formed instead of a spire, and the apex of the shell lies at the centre of this depression. The lower side of the shell is flat. The coiling is sinistral. The aperture is semicircular or oval.

Jurassic (Purbeck) to Recent.

Planorbis is one of the heterogeneous group of fresh-water gastropods widely distributed throughout the world in running or still water. Some can remain submerged; others must surface for air. They feed on algae and plants. They show a wide range in shell form.

Terrestrial gastropod

Helix (fig. 35h) has a thin shell of more or less globular shape; the surface is smooth. There may be an umbilicus. The aperture is obliquely oval.

Eocene to Recent.

Helix is a typical terrestrial gastropod; it is independent of water for breeding, and its mantle cavity acts as a lung. Terrestrial forms are found in a wide variety of habitats throughout the world; they are particularly common in areas where the climate is humid, and where the soil is calcareous.

SUMMARY The typical gastropod has a spirally coiled shell; in forms with a cap-shaped shell, coiling is confined to the apex. The form of the shell is not a reliable guide to the habitat of the animal; however some broad generalisations can be made.

Among marine gastropods: (i) those living in the littoral zone usually have a thick shell; the shell may be cap-shaped as in the limpets, or it may have a rounded shape and a short spire as in the winkles; (ii) forms in which the shell has an entire aperture are often herbivorous, and usually live on a hard substratum; *Natica* and *Turritella* are notable exceptions; (iii) forms with a siphonal canal are often found on soft sediment; and are carnivorous. Fresh-water gastropods for the most part have thin shells with a thick periostracum.

Additional common genera

Gastropods with an exhalant slit

Bellerophon (figs. 37a, 39a) is one of the few gastropods with a bilaterally symmetrical shell; it is smooth and is coiled in a planospiral; each whorl envelops the previous so that only the last whorl is visible. Confusion of *Bellerophon* with a nautiloid is only momentary for it lacks the cephalopod's septa and suture. It is further distinguished by the exhalant slit which cuts the outer lip in the median plane; its earlier trace is visible as a slit band.

Silurian to Lower Trias.

Euomphalus (fig. 37c) has a depressed spire, and a wide umbilicus. The whorl section is polygonal; the slit band forms a spiral ridge between the suture and the periphery.

Silurian to Permian.

Pleurotomaria (fig. 37b) has a moderately high or slightly depressed spire and there may be an umbilicus. The exhalant slit is long, and the slit band forms a spiral angulation on the whorls. The ornament consists of spiral lines and tubercles.

Jurassic to Lower Cretaceous.

Gastropods with a siphonal canal

Fusinus (fig. 38c) has a spindle-shaped shell with many whorls; the siphonal canal is long and narrow. The ornament may be strong; it consists of spiral and transverse ribs. Modern forms live in subtropical waters.

Cretaceous to Recent.

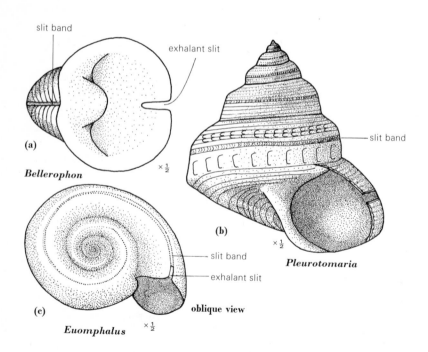

37   Gastropods with an exhalant slit.

Athleta (fig. 38b) has a short spire and a large last whorl. The aperture is long and narrow; the siphonal canal is short. There are spiral markings and callus on the inner lip. The ornament consists of spirally arranged tubercles which are prominent on the posterior part of the last whorl.

Cretaceous to Recent.

Neptunea (fig. 38a) resembles *Buccinum* (fig. 32a), but the spire is more elongate and the siphonal canal wider. The shell is smooth or may show transverse growth lines or spiral ribbing. One common species shows sinistral coiling.

Eocene to Recent.

Fresh-water gastropod

Viviparus (figs. 11, 35g) has a smooth shell with a thick periostracum; the spire is high with a rounded apex; the aperture is entire.

Jurassic to Recent.

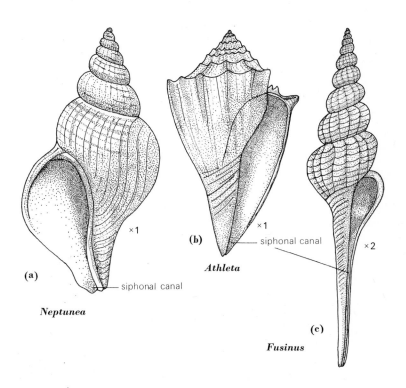

× 1

(b)

siphonal canal

× 1

Athleta

(a)

siphonal canal

× 2

Neptunea

(c)

Fusinus

38   Gastropods with a siphonal canal.

39 Palaeozoic gastropods.

Left, *Bellerophon*, Lower Carboniferous, showing the slit band (×2). Right, *Poleumita*, Wenlock Limestone, Silurian, a form related to *Euomphalus* but differs in having a relatively pronounced ornament (×1).

Geological history of the gastropods

SUMMARY Gastropods are first recorded in early Cambrian rocks. Many families appeared in the course of the Palaeozoic and Mesozoic; few of these became extinct and by the Tertiary gastropods were numerous and highly diversified. Today gastropods far outnumber any other class of mollusc.

The range of shell shape found in Palaeozoic gastropods includes both cap-shaped and spirally coiled shells; amongst the latter planospiral shells were prominent. In most shells the aperture was entire but many forms possessed an exhalant slit, for instance *Bellerophon* (fig. 39) and *Euomphalus* (fig. 37c). Gastropods are found in various types of deposits; they are however commoner in limestones such as the Wenlock Limestone and the reef limestones of the Lower Carboniferous.

In most Mesozoic gastropods the aperture is entire, but forms with a siphonal canal made their appearance in the early Mesozoic and became increasingly important during the Cretaceous. Various Palaeozoic families survived; pleurotomariids (fig. 37b) for example which persist to the present day, living in warm seas. Fresh-water gastropods, too, made their appearance during the Mesozoic, e.g. *Planorbis* and *Viviparus* (fig. 35). These closely resemble living species and can be accepted as fresh-water inhabitants with reasonable confidence.

The dominant Tertiary gastropods possessed a siphonal canal although forms with an entire aperture remained common. A number of genera occurring in the earlier Tertiary deposits still survive and their present-day distribution in warm water supports evidence of other fossils on changes in climate since the Tertiary. *Fusinus* and *Athleta* (fig. 38) are examples common in marine Eocene rocks (e.g. in the Barton Beds) which today live in subtropical seas. Fresh-water and land gastropods are common in some horizons, e.g. in the Bembridge Limestone.

In the Plio–Pleistocene, gastropods are prominent in the Crag deposits. In the earlier beds there are various 'warm' water types but 'cold' water species appear in the later beds. Some, like *Buccinum undatum* (fig. 32a), are still found around the coasts of Britain; a related genus, *Neptunea* (fig. 38a), is today confined to northern waters (see also, fig. 13a).

PTEROPODS

The pteropods, sea-butterflies, are small pelagic gastropods; they are abundant in the surface waters of the ocean and feed on smaller planktonic organisms. They secrete fragile, conical or spiral shells. These may accumulate on the sea-floor in moderately shallow water as pteropod ooze; in greater depth of water the shells are dissolved before reaching the bottom. Pteropods appear first in Tertiary rocks.

TECHNICAL TERMS

APERTURE the opening of the shell through which the head and foot extend (fig. 32a).

APEX the first-formed part of the shell at the tip of the spire (fig. 32a).

CALLUS a subsequent layer of shell over the inner lip and adjoining part of the last whorl (fig. 32a).

COLUMELLA the solid axis of the spire formed by the inner wall of the coiled shell (fig. 32a).

EXHALANT SLIT a narrow fissure cutting the outer lip, through which faeces is passed away from the aperture (fig. 37).

HELICOCONE the distally widening spirally coiled tube which constitutes the typical gastropod shell.

OPERCULUM the lid, horny or calcareous, which closes the aperture when the head and foot are withdrawn into the shell (fig. 32b).

OUTER LIP the outer edge of the aperture; the inner margin is the inner
   lip (fig. 32a).
SIPHONAL CANAL the gutter-like extension of the anterior end of the
   aperture which is moulded around the inhalant siphon (fig. 32a).
SLIT BAND the trace of the exhalant slit filled in by shell during growth
   (fig. 37).
SPIRE all the whorls except the last whorl (fig. 32a).
SUTURE the spiral line along which the whorls are contiguous (fig. 32a).
UMBILICUS the cavity around the axis of coiling found in loosely coiled
   shells (fig. 35d).
WHORL a complete turn of a coiled shell; the ultimate one is the last
   whorl (fig. 32a).

8

Mollusca, class Cephalopoda

The cephalopods are distinguished by the 8, 10 or many prehensile arms, or tentacles which surround the head. The foot is modified and in part forms the FUNNEL, a swimming organ powered by water ejected from the mantle cavity. One living, and most fossil cephalopods possess an external univalve shell. This may be straight, curved or coiled in a plane spiral, and it is divided into chambers by transverse partitions, SEPTA; the septa are perforated by a tube, the SIPHUNCLE which extends from the mantle to the apex of the shell. In most living and some fossil cephalopods the shell is either modified to an internal structure, or it is absent (*Octopus*). In this respect then, living cephalopods are untypical of the fossil record of the class. Cephalopods are the most highly organised molluscs. They include the largest and the fastest-moving invertebrates. Numerically they are long past their zenith.

Living cephalopods are exclusively marine. They range from shallow to deep water; some live at abyssal depths. They are predatory creatures of diverse habit. They include the torpedo-like, fast-moving squids (some can glide for short distances out of water), the less actively swimming cuttlefish, the relatively sedentary *Octopus* which lurks under rocks, and *Nautilus*. *Nautilus* which alone has an external shell swims gently near the sea-floor.

The cephalopods which have an extensive fossil record are separated into the following subclasses:

Nautiloidea: Upper Cambrian to Recent.
Ammonoidea: Devonian to Cretaceous.
Coleoidea, order Belemnoidea. Upper Carboniferous to Eocene.

NAUTILOIDEA

The nautiloid shell is divided into chambers by saucer-like septa with circular or gently fluted edges. Each septum has a more or less central perforation for the siphuncle and this is surrounded on the posterior (convex) face of the septum by a short tube, the septal neck.

The structure of fossil nautiloids is interpreted by reference to the single living genus, *Nautilus* (fig. 40a). *Nautilus* lives in the last formed chamber, the BODY CHAMBER, at the anterior end of its shell. The body can be retracted within this chamber and the aperture closed by a muscular hood. The shell is thus a support and a refuge; it is also, by virtue of gas contained in the chambered part, a hydrostatic organ which facilitates a pelagic life.

Morphology

SOFT BODY. *Nautilus* has a distinct head with a pair of highly developed eyes, and surrounding the mouth there are many retractable tentacles; these lack hooks or suckers that may occur in other cephalopods. Two horny jaws, like a parrot's beak, lie at the opening of the mouth. The body is covered by a thin mantle which secretes and adheres to the shell. On the ventral side of the body the mantle encloses the mantle cavity containing two pairs of gills; all other living cephalopods have only one pair of gills. The mantle is prolonged as a fleshy cord (the siphon), which is enclosed by the siphuncle and this passes back through the chambers to the apex of the shell (fig. 40a).

The FUNNEL is a flexible tube-like structure (fig. 40a). It opens just below the mouth, and its other end leads into the mantle cavity. Oxygenated water drawn in around the edges of the mantle bathes the gills, and is passed out via the funnel; when this water is ejected with force it propels the animal backwards. *Nautilus* alters course by bending its funnel.

SHELL. The shell consists of aragonite with an organic substance (conchiolin). There are two layers, an outer opaque layer with minute grains of aragonite in an organic matrix, and an inner layer composed of nacre (p. 36).

In its simplest form the nautiloid shell is of conical shape; it is closed at the apical end and open at the other end, the aperture. This form of shell is described as an ORTHOCONE (fig. 40g). More generally the shell is either curved (fig. 41), or is coiled in a plane spiral, with the ventral side, the VENTER, forming the 'circumference'. One complete coil is a WHORL, and there may be several whorls. Since the shell widens with growth the last whorl encloses a depression, the UMBILICUS, on each side of the shell, centred on the axis of coiling (fig. 40b). The umbilicus is wide in a loosely coiled, or EVOLUTE shell. In a tightly coiled,

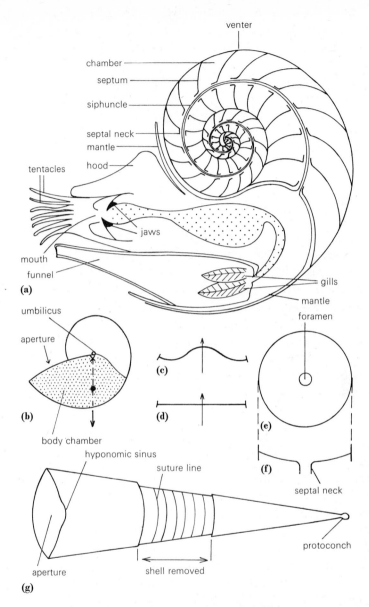

40 Morphology of the nautiloids.

a–c, *Nautilus;* a, a simplified median section of the shell to show the arrangement of the soft parts and internal structures; b, attitude of the shell when floating in water; x is the approximate position of the centre of buoyancy and ● the centre of gravity); c, suture line (the arrow is pointing in the direction of the aperture). d–g, *Orthoceras;* d, suture line; e, anterior view of a septum; f, transverse section of a septum; g, idealised view of a shell showing the main features.

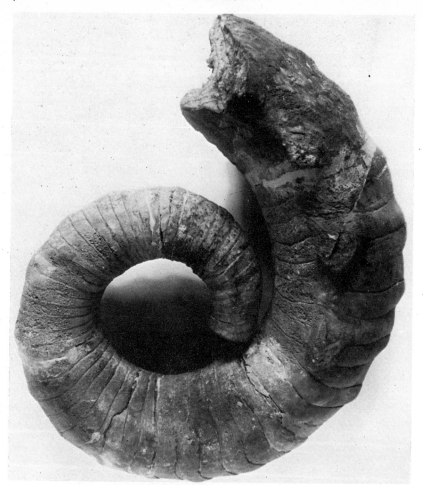

41 A nautiloid with a loosely coiled shell; *Halloceras*, Middle Devonian, Rhineland (×0·6).

INVOLUTE shell, each whorl embraces the previous one closely so that the umbilicus is small.

The chambered part of the shell is called the PHRAGMOCONE. In *Nautilus* the chambers contain a gas which is approximately like air at atmospheric pressure; some chambers contain a little fluid. The gas in the chambers increases the buoyancy of the animal relative to sea-water. The SEPTA are concave forwards (fig. 40a), and they unite with the shell wall so that their edges are seen only by removal of the shell wall. In fossils the septal edges (seen only in internal moulds) form lines, the SUTURE LINES, which are straight in orthocones (fig. 40g, d) and gently

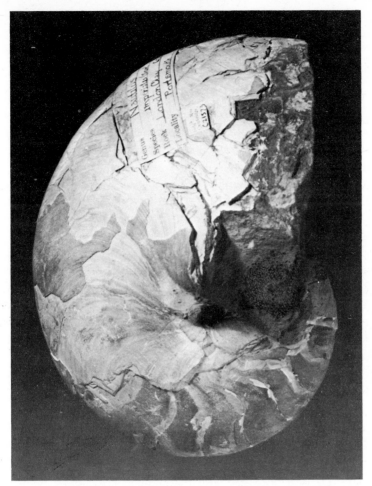

42 *'Nautilus' imperialis*, **London Clay, Eocene** (×1).

undulating in coiled shells (fig. 40b, c); (contrast with ammonite suture lines, pp. 81, 83). Each septum is pierced in the centre by an opening, the FORAMEN (fig. 40e) through which the siphon, enclosed in its shelly tube, the SIPHUNCLE, passes; a short tubular extension of the septum, the SEPTAL NECK, encircles the siphuncle on the posterior side (fig. 40f).

The last chamber is the body chamber and in *Nautilus* it is relatively large; it occupies about one-third of a whorl. The inner surface is smooth apart from some broad muscle scars. The APERTURE varies in shape; it may be round or oval, but sometimes is constricted by inward growth of the margin. In coiled shells there is a re-entrant on the inner (dorsal) side of the aperture due to overlap on the previous whorl; this re-entrant is deeper in involute shells. Most nautiloids have an embayment, the HYPONOMIC SINUS, on the ventral margin of the aperture to accommodate the funnel.

Shell form and mode of life of nautiloids

Nautiloids are abundant and highly varied as fossils, especially in Palaeozoic rocks. Speculation about their behaviour is based largely on a study of the living *Nautilus*. *Nautilus* (figs. 40a–c, 42), has a relatively large smooth shell, commonly being 15 cm or more in diameter. The shell consists of a few whorls coiled in a plane spiral. The last whorl envelops the previous whorls so that the umbilicus is small or absent. The aperture is oval with a deep dorsal re-entrant. There is a hyponomic sinus on the venter. The septa are concave on the anterior side and the suture lines are gently folded. The foramen is central and the short septal neck projects backwards.

Oligocene to Recent. Closely similar genera range back to the Trias.

Nautilus had a wide geographical distribution during the Tertiary; today it is confined to the south-west Pacific. It lives in warm, relatively shallow water, but has been dredged from about 500 m. *Nautilus* is gregarious; it swims in gently moving shoals near the sea-floor and comes inshore at night to feed on small animals (e.g. crustaceans), alive or dead. Its tentacles catch and hold its prey which is cut up by the horny jaws. The shell has cream and brown markings and it is almost invisible in the water. It is beautifully streamlined and it floats in water in a fairly stable position with the aperture turned upwards as in fig. 40b.

Orthoceras (fig. 40d–g). The shell is straight or slightly curved. It increases gently in diameter, and the adult part is almost cylindrical. The cross-section is more or less circular. The septa are concave forwards with the foramen typically in the centre. The suture line is straight. The surface of the shell is smooth.

Middle Ordovician. Many related forms extend the range to the Trias.

Orthoceras and similar orthocones range in length from about 3 cm to possibly 460 cm with a maximum diameter of about 30 cm. The larger shells must have been rather vulnerable to damage in shallow turbulent water; certainly fossils are often incomplete. It is probable that *Orthoceras* and similar forms swam in typical cephalopod fashion by jet propulsion; this is implied by their having a hyponomic sinus. Probably, too, they moved horizontally; certainly 'camouflage' markings, preserved in rare cases, occur on one side only of the shell, the dorsal. Because of

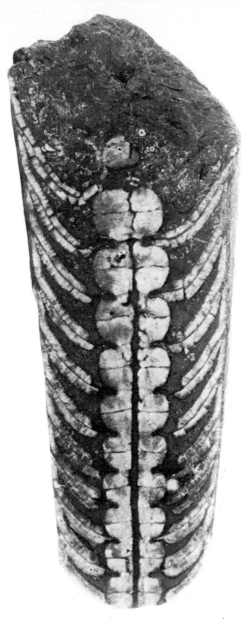

43   A section of a straight-shelled nautiloid, with calcareous deposits on the septa and also in the siphuncle which is expanded between the septa resembling a string of beads; *Rayonnoceras*, Lower Carboniferous (× 0·9). It is estimated that individual specimens of this form may have reached a length of 6 m.

its buoyancy, the shell must have tended to float in a vertical position with the aperture facing downwards. In a number of orthocones, shelly deposits in the chambers and around the siphuncle may have reduced the buoyancy sufficiently to enable the animal to swim with its shell held in a horizontal position (see fig. 43). In this connection, the way in which buoyancy is adjusted in some living cephalopods is interesting; for instance an ingress of water to the cuttlefish bone has been observed, and liquid may occur in some chambers in *Nautilus*.

Geological history of the nautiloids

SUMMARY Nautiloids are the earliest of the cephalopods. They occur first in Upper Cambrian rocks; they are abundant and diversified throughout the Palaeozoic, but they nearly became extinct at the end of the Trias. Numbers increased again for a time during the Mesozoic, but they gradually declined during the Tertiary until at the present day only *Nautilus* remains.

The predominant nautiloids of the Palaeozoic have straight or curved shells; coiled forms are, however, quite common. In some, additional shelly material is formed around the siphuncle and septal necks. Straight-shelled nautiloids disappear at the end of the Trias, leaving only planospiral forms ranging from evolute to involute, and with a wide range in shape of aperture. The more involute forms have sutures which are quite strongly zigzag. *Nautilus* dates from the Oligocene, but very similar genera occur in the Mesozoic.

AMMONOIDEA

In the ammonoids the suture line is strongly folded and often very complex, the septal necks in adult shells project forwards, and the siphuncle typically lies close to the ventral margin of the shell; the embryonic shell (protoconch) is present.

The shell consists of three parts, the PROTOCONCH, the PHRAGMO-CONE and the BODY CHAMBER. In some forms the aperture was closed by one or a pair of plates (APTYCHI). The adult shell is a record of the stages in growth of the shell.

The term 'AMMONOID' includes the GONIATITES (Palaeozoic), the CERATITES (Permo-Trias) and the AMMONITES (Mesozoic). The group is extinct and the nature of the soft body can only be surmised with reference to *Nautilus*, the one surviving cephalopod with an external chambered shell. The differences between the ammonoid and

the nautiloid shells are, however, quite marked, and possibly the soft
parts, too, were not closely similar.

Morphology

The shell is made of aragonite; it is usually thin, and in structure it
resembles *Nautilus*. The aptychi consist of calcite.

The ammonoid shell is typically coiled in a plane spiral (fig.44a);
less commonly the shell may be straight, curved or combine initial
coiling with a distal straight or hooked section; a helicoid spiral shell is
exceptional. Coiled shells may be *evolute* or *involute*; their shape is also
influenced by the form of the cross- or WHORL section which may be
round, oval, quadrate, compressed, triangular or depressed (fig.45). The
whorl section in involute shells is modified by the re-entrant on the

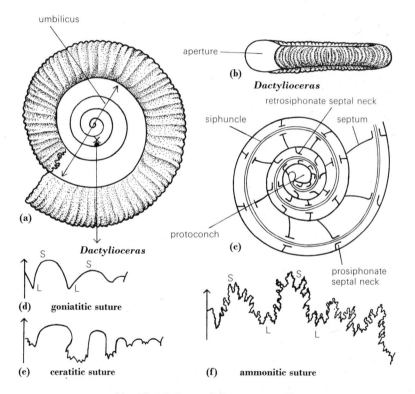

44 Morphology of the ammonoids.

a, b, c, *Dactylioceras;* a, diagram of the shell in its presumed position of life (body chamber,
ornamented; phragmocone, ornament omitted; x, centre of buoyancy; ●, centre of gravity);
b, front view showing the aperture and shape of the whorl section; c, median section through
the early whorls of a shell, much enlarged. d–f, ammonoid suture lines; L, lobe; S, saddle.

45 An involute and highly globose ammonite with a depressed whorl section; *Cadoceras*, Kellaways Rock, Upper Jurassic ($\times 1 \cdot 1$). Above, front view; below, lateral view.

46   Contrasts in ammonitic suture lines.

Left, *Strigoceras*, Middle Jurassic, with a highly intricate suture line. Right, *Hildoceras*,
Lower Jurassic, with a relatively simple suture line. (Both × 0·7.) The area between several
suture lines has been painted to show their course more clearly.

dorsal side which is caused by the outer whorl impinging on the inner
whorl (fig. 44b). Certain combinations of degree of coiling and shape of
whorl section are common, and have been given descriptive terms, e.g.
oxycone, planulate, serpenticone, sphaerocone; these terms are defined
on page 102.

The embryonic shell, or PROTOCONCH is seen only in a median
section of the shell. It is an oval, or barrel-shaped chamber cut off from
the phragmocone by the first septum (fig. 44c).

The PHRAGMOCONE is divided into chambers by septa which
typically are much more complex structures than in the nautiloids. In
all ammonoids the septa are folded where they meet the shell wall, and
the basic plan of their SUTURE (pp. 76–77) is a zigzag line on which the
forward-pointing folds are called SADDLES and the backward folds are
called LOBES (fig. 44d). The folding of the suture line is relatively
shallow in primitive ammonoids, and in the early septa of individuals,
but it may be intensely crimped in the adults of the advanced ammonoids (fig. 46). Suture patterns are of great value in taxonomy, but in
practice their identification is a matter for the specialist; it is sufficient
at this stage to refer a suture line to one of three categories: (i) GONIA-
TITIC suture (fig. 44d) in which both lobes and saddles are entire;

47 A section through an ammonite shell, approximately along the median plane, showing the way in which the shell is preserved, with rock matrix infilling later chambers, and crystalline calcite the earlier chambers; the latter contain cavities into which well-formed calcite crystals project. A part of this section is shown in fig. 48. *Parkinsonia*, Inferior Oolite, Jurassic; diameter of shell, 280 mm.

typical of the goniatites (Palaeozoic); (ii) CERATITIC suture (fig. 44e) in which the saddles are entire but the lobes are toothed; typical of the ceratites (Permian and Triassic); (iii) AMMONITIC suture (fig. 44f) in which both lobes and saddles are intricately frilled; typical of the ammonites (Permian and Mesozoic). Only a part of the suture is usually shown in diagrams, starting from the venter (the position of which is

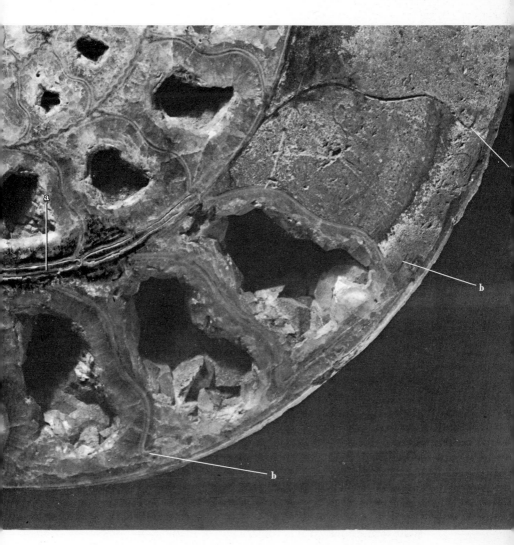

48   The siphuncle seen at a, and septal necks at b, in part of the section shown in fig. 47 (× 1).

49   Aptychi, Kimmeridge Clay, Upper Jurassic (× 1·5).

indicated by an arrow pointing towards the aperture), and ending at the margin of the umbilicus.

The SIPHUNCLE is a slender tube which starts blindly in the centre of the first septum (fig. 44c, see also fig. 48). In most ammonoids it shifts outwards so that in the later whorls it lies immediately under the venter; in one Devonian family it lies on the inner, dorsal margin. Short SEPTAL NECKS encircle the siphuncle where it passes through the septa; the septal necks project forwards (PROSIPHONATE) in the later formed septa of Mesozoic ammonites (figs. 44c, 48) but in the earliest ammonoids (as also in the nautiloids) they project backwards (RETRO-SIPHONATE, see fig. 40a).

The BODY CHAMBER varies in length; it may occupy only half a whorl in stout shells, or more than one whorl in slender shells. It is not always preserved; it lacks the support afforded the phragmocone by the 'fan-vaulting' of the septa and was, perhaps, the more easily broken.

The aperture is, in general, of similar shape to the whorl section; in some forms however it is constricted, or modified by projections of shell. Lateral projections (one on each side) are LAPPETS (fig. 56) and a ventral projection is a ROSTRUM. Two calcite plates, APTYCHI (fig. 49) or a single horny plate are sometimes associated with ammonite shells in Mesozoic rocks and, in rare instances, close the aperture as does the operculum in gastropods. They did not articulate with the shell, and ordinarily must have dropped away from the buoyant shell on decay of the soft parts.

ORNAMENT. The shell may be smooth, or, especially in Mesozoic forms, it may be ornamented by fine lines (striae), ribs, tubercles or spines. Ornament may be transverse, spiral or both transverse and spiral. It may be confined to the sides, or may also occur on the venter. Some forms may have a groove (SULCUS) (fig. 53b) on the venter, or a ridge (KEEL) (fig. 52b). The ornament may be a localised thickening of the shell, but more generally it is a folding of the shell so that it is seen also in internal moulds.

ORIENTATION. Ammonoids are, with few exceptions, bilaterally symmetrical; the plane of symmetry lies at right angles to the axis of coiling. The shell is usually figured in two positions, one showing the side view, and the other the view of the aperture. The periphery is the ventral side.

Shell form and possible attitude in life of some ammonoids

Dactylioceras (fig. 44a, b) has an evolute shell with a wide, shallow umbilicus; the whorl section is oval. It is a typical serpenticone (p. 102). The ornament consists of narrow, closely spaced ribs, most of which bifurcate; they extend across the venter. The suture is ammonitic.

Upper Lias.

The possible life attitude of some ammonoid shells has been investigated using an estimated density of the body (with reference to *Nautilus*) and buoyancy of the shell. In a form like *Nautilus* with an involute shell and a short stout body, the estimated centres of gravity and of buoyancy are relatively widely separated. Such shells are fairly stable in water and float with the aperture facing upwards (fig. 40b). In an evolute form like *Dactylioceras* with a long slender body occupying about a whorl, the centres of gravity and buoyancy almost coincide (fig. 44a). For this type of shell a wider variety of attitudes is possible and *Dactylioceras* could possibly rotate its shell in a vertical plane; thus it may have swum with the aperture facing upwards or crawled, using its tentacles, with the aperture facing downwards. The strong ribbing must have impeded fast movement.

Oxynoticeras (fig. 50a, c) has a compressed involute shell with a small shallow umbilicus. The whorl section is acutely triangular, with a deep re-entrant on the dorsal side. The shell is a typical oxycone (p. 102). The ornament of gently curved ribbing, present on earlier whorls, is almost smoothed out in the adult shell. The suture is ammonitic.

Lower Lias.

Estimations available for a similar shell show that the centres of gravity and of buoyancy in *Oxynoticeras* probably lay as indicated in fig. 50a. These positions compare closely with those in *Nautilus*. *Oxynoticeras* may then have lived with its shell fairly stable in water and with the aperture facing upwards. The sharp 'cut-water' venter is usually interpreted as a design for speed; and *Oxynoticeras* was probably able to move quite rapidly.

Goniatites (fig. 50b, e). The shell is globose and very involute; the whorl section is depressed, and rounded on the ventral side with a wide

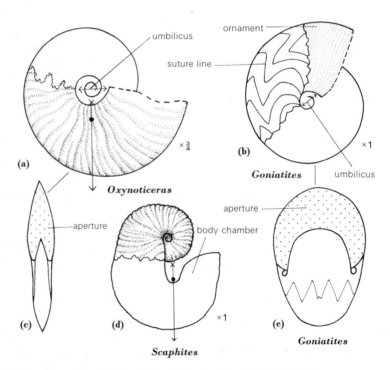

50 The position of the ammonoid shell in life.

a, c, *Oxynoticeras; a,* an oxycone shell in its presumed attitude of life (body chamber ornamented); c, front view showing the shape of the whorl section. b, e, *Goniatites;* b, lateral view with part of the shell removed to show the suture lines; e, front view showing the depressed whorl section. d, *Scaphites* in its presumed position of life.

re-entrant on the dorsal side. The umbilicus is small. The surface of the shell is faintly striated. The suture is goniatitic.

Lower Carboniferous.

Goniatites is an example of a sphaerocone; its shape contrasts strongly with that of *Dactylioceras* and *Oxynoticeras*. Sphaerocones occur repeatedly during ammonoid history, and some of the Mesozoic examples had coarse radial ribbing which must have further impeded movement through the water.

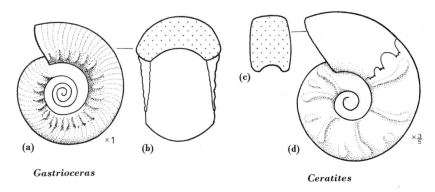

Gastrioceras

Ceratites

51   A goniatite and a ceratite.
a, b, *Gastrioceras;* a, lateral view; b, front view showing the shape of the whorl section. c, d, *Ceratites;* c, whorl section; d, lateral view.

Scaphites (fig. 50d). The shell is moderately inflated; the early whorls are involute and the body chamber (in the adult) forms a short, straight shaft which ends in a hook. The aperture is slightly constricted. The suture is ammonitic.

Upper Albian to Upper Cretaceous.

Scaphites is one of a diversity of ammonites in which the shell, partly or wholly, was either straight, or only loosely coiled; they are referred to as 'uncoiled' ammonites. They are common as fossils, especially in the Cretaceous, and some are relatively long ranged; these points indicate the success of their adaptation to their environment.

The probable attitude of *Scaphites* in life is shown in fig. 50d. This position, with the aperture facing upwards, is thought to have been fairly stable, and appears more suited to floating passively than active swimming. It is possible that *Scaphites* and similar uncoiled varieties floated near the surface of the sea, or by adjusting their density moved up and down in the water.

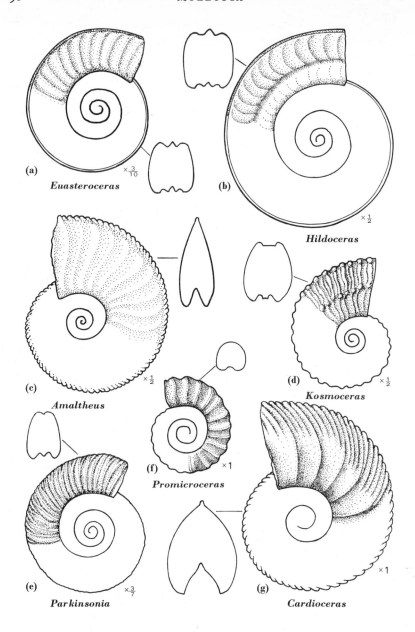

52 Jurassic ammonites.

Various planktonic marine animals live passively in water with their mouths directed upwards. They feed on small organisms caught by their tentacles or, in some cases, on microscopic plankton trapped by ciliary action.

SUMMARY The ammonoid shell was a hydrostatic organ of a design which in most cases facilitated a pelagic life, either actively nektonic or passively planktonic.

The four genera described above are a small sample of the range in shell form found in ammonoids. Each represents a pattern of shell, variations on which recur throughout ammonoid history, and may be presumed to be an adaptation to a particular mode of life. The abundance of fossils of each type on repeated occasions during geological time, is an indication that each was well adapted to its environment.

Additional common genera

Palaeozoic

Gastrioceras (fig. 51a,b, also fig. 2) has a moderately large and deep umbilicus. There are stout tubercles on the margin of the umbilicus from which very fine ribs run across the venter. The suture is goniatitic. Upper Carboniferous.

Gastrioceras is one of the goniatites which occurs in marine shales in Millstone Grit facies, and is useful for zonal purposes, being characteristic of zone G. Other forms are *Eumorphoceras* (zone E), *Homoceras* (zone H) and *Reticuloceras* (zone R, fig. 54). They are relatively involute forms distinguished from each other mainly by their ornament.

Mesozoic

Ceratites (fig. 51c,d). The shell is relatively compressed and evolute with a shallow umbilicus. The ornament consists of broad ribs with tubercles. The suture line is ceratitic. Trias (Muschelkalk).

The following genera have an ammonitic suture line:

Euasteroceras (fig. 52a) has a moderately evolute shell (planulate) with a wide shallow umbilicus. On the venter is a strong keel with a shallow sulcus on each side. There are strong radial ribs which curve forwards gently on the shoulders. Lower Lias.

Promicroceras (fig. 52f) has a small evolute shell with strong radial ribs which cross the venter.
Lower Lias.

Amaltheus (fig. 52c) has a compressed involute shell (oxycone). The ornament consists of ribs which bend forwards towards the keeled venter; extra ribs on the keel give it a crenulated appearance.
Middle Lias.

Hildoceras (fig. 52b) has a relatively evolute shell with a slightly compressed quadrilateral whorl section; there is a keel bordered on each side by a sulcus. The ornament consists of sickle-shaped ribs on the sides.
Upper Lias.

Parkinsonia (fig. 52e) has a relatively evolute (planulate) shell. The ornament consists of sharp ribs which bifurcate towards the venter; the ribs are interrupted on the venter by a sulcus.
Inferior Oolite.

Kosmoceras (fig. 52d) has a compressed shell with a flattened venter. It is moderately evolute. There are relatively sharp ribs which bifurcate; tubercles occur where the ribs split and also along the margins of the venter. Lappets are developed (fig. 56).
Oxford Clay.

Cardioceras (fig. 52g) has a moderately evolute and compressed shell. Strong curved ribs, with shorter ribs intercalated, extend across the keeled venter.
Oxford Clay.

Euhoplites (fig. 53b) has a compressed shell with a sulcus on the venter. It is moderately evolute. Prominent ribs arise from tubercles on the umbilical margin, and curve forwards to join more prominent tubercles which border the sulcus.
Lower Gault.

Schloenbachia (figs. 53a, 57) has a relatively compressed shell with a keeled venter; it is moderately evolute. There are strong curved ribs on the sides, with tubercles at intervals.
Lower Chalk (Chalk Marl).

Hamites (fig. 53d). The early part of the shell is coiled, but the main part is bent to form three subparallel shafts. The whorl section is oval or rounded. The ornament consists of strong, closely spaced ribs.
Gault.

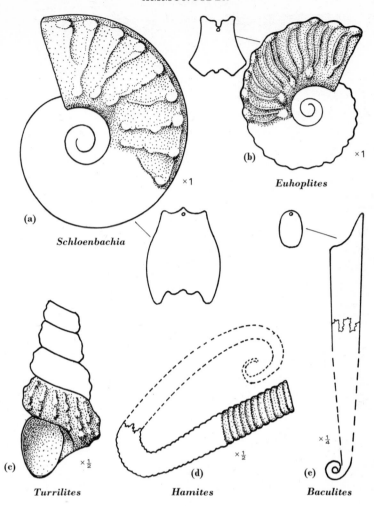

(b)

×1

Euhoplites

×1

(a)

Schloenbachia

(c) ×½

Turrilites

(d) ×½

Hamites

(e) ×¼

Baculites

53   Cretaceous ammonites.

Turrilites (fig. 53c) has its shell coiled in a helicoid spiral. Cenomanian.

Baculites (fig. 53e). The shell is straight except for the apical part which is coiled. The whorl section is oval.
Middle and Upper Chalk.

Geological history of the ammonoids

SUMMARY The ammonoids had separated from nautiloids by the Devonian. The number of genera remained fairly small until the Trias

54   A goniatite from shales in the Millstone Grit Series, Upper Carboniferous.
Reticuloceras, Zone R (×8).

when a marked proliferation and diversification took place. They almost died out at the end of the Trias, but a resurgence of numbers in the Jurassic culminated in the maximum development of the group. Their numbers then declined slowly until their extinction at the close of the Cretaceous.

Palaeozoic ammonoids are commonly spoken of as 'goniatites'; the term is descriptive of the angular nature of the goniatitic suture line which they display. The goniatites are typically small, more or less involute and smooth shelled; a strong ornament is exceptional. *Gonia-*

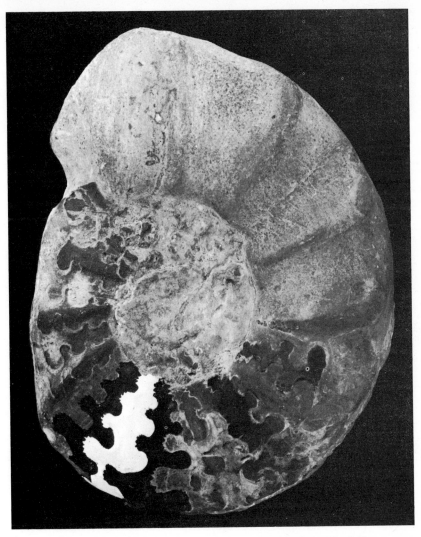

55   A Triassic ammonoid, *Ceratites*, Middle Trias (×1·3).

tites (fig. 50b) is found in shales, and sometimes limestones in the Lower Carboniferous. Other goniatites are used in the correlation of the Millstone Grit facies; they occur as crushed specimens in black shales, or as uncrushed fossils in nodules (bullions fig. 54). Few other fossils are associated with these goniatites apart from bivalves (*Dunbarella* (fig. 29a) and *Posidonia*).

Goniatites survived into the Permian but they are outnumbered there

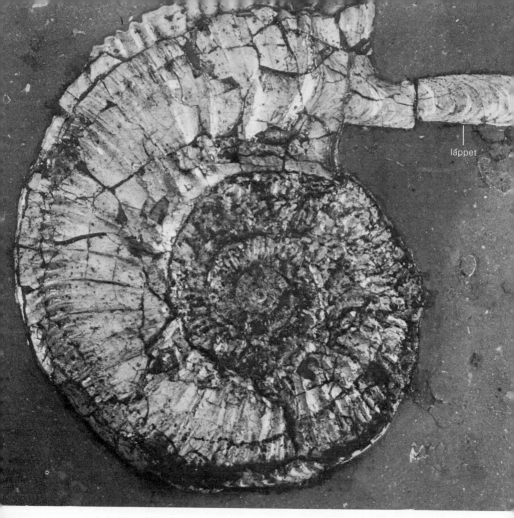

läppet

56   A Jurassic ammonite. *Kosmoceras*, Oxford Clay ($\times 1\cdot7$).

by forms with a more complex suture. The most important of these had a ceratitic suture, and were pre-eminent until the Middle Trias; *Ceratites* (Middle Trias, fig. 55) is the best known example. True ammonites, with an ammonitic suture appeared in the late Palaeozoic but they occupied a subordinate position until the Upper Trias when their numbers greatly increased; during this time they reached an acme of diversity and the most intricate suture line known is found in a Triassic ammonite.

Ammonoids are generally widespread in most marine rocks of Permo-Trias age in Europe and North America. They are absent from Britain

57   A Cretaceous ammonite. *Schloenbachia varians*, **Chalk Marl, Lower Chalk (×3).**

where the only marine sequence of this time, the Magnesian Limestone, represents conditions unfavourable to ammonoids.

Only a single family survived from the Trias into the Jurassic. (*Monophyllites*, Plate IV, is a primitive member), and from it a succession of forms was derived, directly and indirectly. Their diversity and the vast numbers in which they were preserved has greatly simplified the zoning and correlation of the Jurassic, and to a lesser extent the Cretaceous rocks. Their usefulness stems from a combination of factors: (i) they are typically well preserved and common as fossils; (ii) they are relatively easy to distinguish; (iii) they occur widely; (iv) they appear in relatively quick sequence; and (v) they occur in a range of facies.

BELEMNOIDEA

The belemnites (or belemnoids) are extinct cephalopods whose shell was enclosed by the soft body. It consisted of a phragmocone (with protoconch) partly contained in a cavity in one end of a solid calcareous guard.

Morphology

The GUARD is normally the only part preserved of the belemnite shell. It is a cigar-shaped solid structure, tapered at the posterior end, and with a deep conical cavity, the ALVEOLUS, at the anterior end (fig. 58a, e). In most genera, the guard is not a large structure, falling within a range of 2–20 cm in length. In transverse section, the guard shows a fibrous structure with tiny calcite prisms radiating from an eccentric point (fig. 58c); rings concentric about this point denote growth stages. The PHRAGMOCONE (figs. 58a, 59) is less commonly preserved. It is a conical, chambered shell with thin delicate walls. The posterior end,

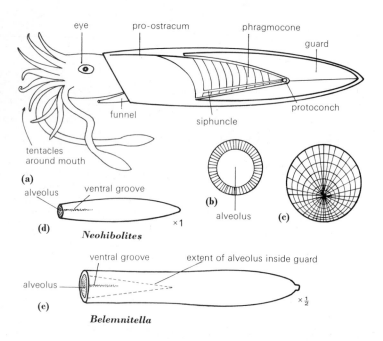

58   Morphology of the belemnites.

a, simplified section of the skeleton of a belemnite in the plane of bilateral symmetry; a possible arrangement of soft parts is sketched in. b, c, transverse sections across the guard in the region of the alveolus (b), and below the alveolus (c). d, e, examples of belemnites.

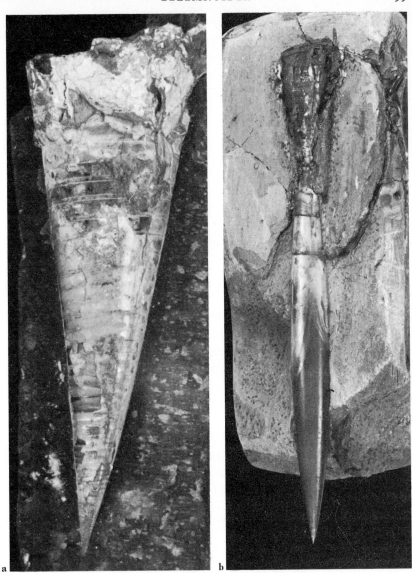

59   Belemnites with the phragmocone preserved.

a, a guard, fractured lengthwise along the median plane, with a part of the uncrushed phragmocone lying *in situ* in the alveolus; *Belemnites oweni*, Oxford Clay, Jurassic; (×2). b, a
guard with the anterior part of the phragmocone (crushed); *Neohibolites minimus*, Gault
Clay, Cretaceous (×2).

a

pro-ostracum

rows of hooklets

b

60   A belemnite with traces of arms.
a, the paired rows of hooklets indicate the traces of at least eight arms (×0·7). b, hooklets
(×1·3). *'Belemnoteuthis'*, Lias, Lower Jurassic, Lyme Regis.

with the oval protoconch at its tip, lies within the alveolus of the guard.
There is no body chamber, and the anterior end is drawn out dorsally
as a fragile horny process, the PRO-OSTRACUM. The septa are saucer-
like with unfolded edges. The siphuncle lies nearer the ventral margin
of the phragmocone.

A common example of a belemnite

Neohibolites (figs. 58d, 59b). The guard is small, usually about 2–8 cm
in length. It is more or less spindle-shaped, tapering to an acute point
at the apex (posterior end). The alveolar region is not usually preserved,
but in specimens where a part of it remains there is a short ventral
groove.
    Gault.

Mode of life of the belemnites

The belemnites have left a substantial fossil record in Mesozoic rocks
which connects them indirectly with the modern two-gilled forms like
the cuttlefish and squids which are included in the order Decapoda.
These forms have ten suckered arms round the mouth, and their shell is
internal and much modified compared with the belemnite shell. The
cuttlefish shell or 'bone' consists mainly of a wide flat phragmocone
with obliquely overlapping septa, and with a tiny remnant of guard at

the posterior tip. Decapods are streamlined animals and the squid, in particular, can move at very high speed. They possess an ink sac from which a cloud of dark pigment can be ejected through the funnel to cover their retreat from danger.

Exceptionally, fossil belemnites may show traces of the soft parts, including impressions of the arms with double rows of horny hooks (see fig. 60), and also the ink sac.

Belemnites presumably swam in similar fashion to the modern decapods, with the body held in a more or less horizontal position. The guard, relatively heavy and rigid at the posterior end, would counterbalance the buoyancy of the phragmocone.

Geological history of the belemnites

SUMMARY Belemnites first appear in rocks of Upper Carboniferous age. They range generally to the end of the Cretaceous but in one area they persist into the Eocene.

The belemnites are most abundant and diverse in the Jurassic and Cretaceous. Their remains occur in a variety of rock type, but they are rather more common in clays. There are many genera of belemnites, distinguished mainly by the shape of the guard, and the arrangement of surface grooves. They are of some value stratigraphically; for instance, they provide two zonal indices in the Chalk.

TECHNICAL TERMS

ALVEOLUS the cavity in the anterior end of the guard within which the phragmocone is located in belemnites (fig. 58e).

APERTURE open end of body chamber (fig. 44b).

APTYCHI a pair of plates which closed the aperture of some ammonites, (fig. 49). Some forms had a single horny plate, the ANAPTYCHUS.

BODY CHAMBER the last-formed chamber which contained the body.

CHAMBER (or camera) space enclosed by two adjacent septa (fig. 40a).

EVOLUTE term describing a loosely coiled shell in which the outer whorls enclose the inner whorls to only a small extent; evolute shells have a wide umbilicus (fig. 44a).

FORAMEN the hole in the septum through which the siphuncle passes (fig. 40e).

FUNNEL the tubular swimming organ through which water is ejected from the mantle (fig. 40a).

GUARD cigar-shaped structure enclosing the posterior end of the phragmocone in belemnites (fig. 58a).

HYPONOMIC SINUS a curved re-entrant on the ventral margin of the aperture for the passage of the funnel (or hyponome) in nautiloids (fig. 40g).

INVOLUTE refers to a tightly coiled shell in which the inner whorls are largely concealed by the last whorl; the umbilicus is small in involute shells (fig. 50b).

KEEL a ridge on the venter (fig. 52b).

LAPPET a forward projection of shell on each side of the aperture in some ammonites (fig. 56).

ORTHOCONE a 'straight' nautiloid shell (fig. 40g).

OXYCONE an involute shell with a sharp venter (fig. 50a, c).

PHRAGMOCONE the chambered part of the shell (figs. 40a, 44c).

PLANULATE describes an evolute shell with a more or less oval whorl section (fig. 52e).

PRO-OSTRACUM a dorsal projection of the phragmocone in belemnites (fig. 58a).

PROTOCONCH the initial part of the shell occupied by the embryo (fig. 44c).

ROSTRUM a forward projection of shell on the ventral side of the aperture in some ammonites.

SEPTA (septum) transverse partitions, which divide the phragmocone into chambers (figs. 40a, 44c).

SEPTAL NECK a funnel-like projection of the septum around the siphuncle (fig. 40a, f).

SERPENTICONE describes an evolute shell with many slender whorls which barely overlap (fig. 44a).

SIPHUNCLE a long slender tube extending from the body chamber back to the protoconch (figs. 40a, 44c).

SPHAEROCONE an almost spherical, very involute shell with a very small umbilicus and a depressed whorl section (fig. 50b, e).

SULCUS a groove on the venter (fig. 53b).

SUTURE LINE the trace of the edge of a septum exposed by removal of the shell wall; it is seen only in internal moulds (fig. 46). The suture line may be simple or folded; a fold convex towards the aperture is called a SADDLE, and a fold concave towards the aperture is a LOBE.

TYPES OF SUTURE LINE (i) SIMPLE, a straight line when projected on paper; e.g. *Orthoceras* (fig. 40d); (ii) GONIATITIC, the suture line is bent into strong lobes and saddles: e.g. *Goniatites* (fig. 44d); (iii) CERATITIC, the saddles are rounded, and the lobes are toothed, e.g. *Ceratites* (fig. 44e); (iv) AMMONITIC, both saddles and lobes are strongly frilled, e.g. most Mesozoic ammonites (fig. 44f).

UMBILICUS the depression enclosed by the last complete whorl on each side of a planospiral shell (fig. 44a).

VENTER the ventral wall of a coiled shell, usually on the outside (fig. 40a).

WHORL one complete coil of a shell.

9

Echinodermata, class Echinoidea

The echinoderms, which are exclusively marine animals, have a long fossil record ranging back to the early Palaeozoic. The main classes common as fossils are the Echinoidea (sea-urchins), Crinoidea (crinoids) and Stelleroidea (starfish and brittle stars), all of which have living representatives; and the Cystoidea and Blastoidea, which were common in the Palaeozoic but are now extinct.

Echinoderms have a number of unique features. Their skeleton (test), which is internal, consists of a variable number of small plates, each of which is a single calcite crystal. Within the test is a hydraulic system of tubes, the WATER VASCULAR SYSTEM, along which water is circulated in the body. Most forms have a five-rayed arrangement of the body parts as seen, for instance, in the starfish, and the basic symmetry is bilateral. But, where the five rays are regularly spaced, the symmetry is radial.

Internal organs include a gut, with mouth and anus, but there is no head, heart or special excretory system. The nervous system is a network of nerve cords following the plan of symmetry of the body, and sense organs are poorly developed. The various organs are enclosed in the body cavity, or coelom, and an extension from this forms the WATER VASCULAR SYSTEM. This consists of a series of tubes through which fluid, mainly sea-water, is circulated in the body. Sea-water enters the system via a porous plate, the MADREPORITE, and is distributed along radially disposed tubes, the RADIAL CANALS (fig. 61j), to soft tentacles, the TUBE FEET, which extend from the surface of the body and which are located in five bands, the AMBULACRA, overlying the radial canals. The tube feet are actuated by variation of hydrostatic pressure.

In mode of life, most echinoderms are benthonic, though a small number are pelagic, and range from shallow water to abyssal depths, wherever normal salinity occurs. The echinoids have locomotory organs and may move freely in search of food; many of them have jaws with which to macerate this. The stelleroids, too, are vagrant. The crinoids, cystoids and blastoids, however, are (or were) mostly sessile forms, feeding passively on microscopic food collected by ciliary currents.

In echinoderms, the sexes are separate and fertilisation of the eggs takes place in the sea. The larvae swim freely among the plankton for a time, becoming widely dispersed in the process.

The echinoids, which are abundant in Mesozoic and Cainozoic rocks and which are often stratigraphically important, will be considered in more detail than the other classes.

CLASS ECHINOIDEA

The echinoid test is a rigid structure which may be hemispherical, disc-shaped or heart-shaped, and consists of many interlocking plates which are covered by skin. It is, therefore, an ENDOSKELETON. The outer surface is covered by spines, which serve as a protection and which can also be used for walking. There are no arms.

Echinoids are gregarious benthonic forms, vagrant to some extent or burrowing in sediment. They live for the most part in shallow coastal waters. Common living examples include the so-called edible sea-urchin, *Echinus*, and the heart urchin, *Echinocardium*.

Morphology

The test of a typical echinoid is hemispherical in shape. It consists of many interlocking plates, arranged in 10 double columns, which radiate from the apex of the upper surface to the mouth in the centre of the lower surface. Five of these columns carry tube feet and are known as AMBULACRA. The other five, with no tube feet, are known as INTER-AMBULACRA. The 10 columns together make up the CORONA. Situated at the apex of the test is the APICAL SYSTEM consisting of about 10 small plates which are connected with specialised functions, and one of which is the MADREPORITE. The central part of the apical system consists of a membrane, the PERIPROCT, which surrounds the anus. In the centre of the lower surface of the test is a similar membrane, the PERISTOME, which surrounds the mouth. Most of the surface of the test is covered by SPINES which are attached to TUBERCLES.

SOFT BODY. The test is filled with fluid in which the rather insubstantial soft organs are suspended (fig. 61j), the gut being looped around

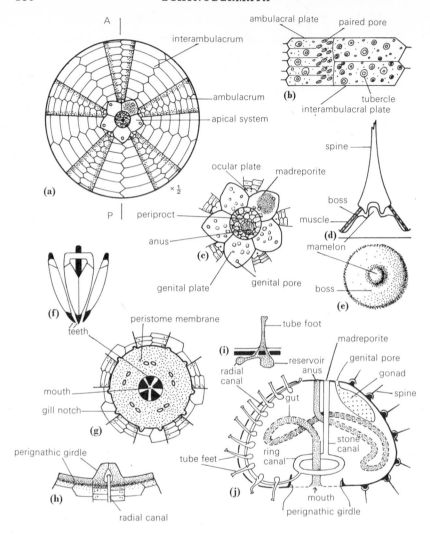

61   Morphology of a regular echinoid, based mainly on *Echinus*.
a, aboral view (A, anterior; P, posterior; line A–P, plane of symmetry). b, ambulacral and
interambulacral plates, enlarged. c, view of apical system, enlarged. d, section through a
tubercle and spine. e, plan of a tubercle. f, side view of the jaws showing a part of the complex
of plates which holds the five teeth (shaded black). g, oral view showing the peristome membrane (stippled) surrounding the mouth. h, a part of the perignathic girdle as seen from inside
the test. i, a section through a tube foot to show its relationship with the paired pores of an
ambulacral plate (thick black line). j, a simplified section through an echinoid test to show the
arrangement of the internal organs (on the left the section cuts an ambulacrum and on the
right it cuts an interambulacrum).

inside the body wall as it ascends from mouth to anus. The radial canals of the water vascular system underlie the ambulacral areas, and the tube feet penetrate through paired PORES in the ambulacral plates to the exterior. These tube feet, typically, end in suckered discs, with which the echinoid can cling to a firm surface. In use, the tube foot is extended by water forced into it from a reservoir (fig.61i) and exerts a suction grip on an object which it touches, by the withdrawal of water. In this way the tube feet may be used either for clinging or in locomotion. In some echinoids, they may also be modified for other functions which may be sensory, for use in feeding, for the construction of burrows, or they may form leaf-like respiratory structures. Some forms carry gills, which occur as bushy outgrowths around the peristome. There are five gonads and their products are discharged through pores in five plates (genital plates) of the apical system.

The outside of the test is covered by skin, and the spines are held in position, and moved, by muscles which are attached to the outer side of the plates of the test. The plates have a network structure with microscopic pores which are permeated with soft tissue. After the death of the animal all the soft body decays and, during fossilisation, the minute pores are normally closed by precipitation of calcite. Since each plate and spine is a single crystal of calcite, these will show the characteristic calcite cleavage when the fossil is broken.

ORIENTATION. When viewed from the upper surface, the outline of the type of echinoid test just described is a circle, with the apical system at its centre, and the double columns of plates forming radii (fig.61a). The ambulacra are said to be RADIAL in position, since each overlies a radial canal of the water vascular system (fig. 61j) and the interambulacra are said to be INTER-RADIAL. The only asymmetric feature in the radial symmetry of this test is the position of the madreporite, and this is used to define the ANTERIOR–POSTERIOR orientation. The test is conventionally aligned as shown in fig. 61a, c, with the madreporite on the right, towards the anterior end. The lower surface, on which the mouth lies, is the ORAL surface; and the upper surface is the ABORAL surface.

TEST. The pattern of symmetry shown by the test is a convenient basis for separating the echinoids into two groups, REGULAR forms, in which the coronal plates show RADIAL symmetry (fig.61a), and IRREGULAR forms, in which the five rays are arranged with BILATERAL symmetry (fig.67a,c). In regular forms the anus (surrounded by the periproct) lies WITHIN the apical system (fig.61a), and the mouth is at the centre of the oral surface and contains jaws. In irregular forms, the anus lies

OUTSIDE the apical system in the POSTERIOR interambulacrum (fig. 73a), and the mouth may lie either in the centre of the oral surface and have jaws (fig. 71l), or towards the anterior margin and lack jaws (fig. 73g).

The plates of the apical system and of the corona may be preserved intact in a fossil echinoid. The positions of the periproct and the peristome, however, are usually indicated by a space.

APICAL SYSTEM. In regular echinoids (fig. 61c), the apical system contains 10 plates, arranged in one or two rings around the periproct. Five of these, the OCULAR plates, are situated radially; and alternating with them are five inter-radial GENITAL plates. The ocular plates each bear a pore through which passes the terminal tentacle of the radial water vessel. The genital plates are the larger and each has a pore through which eggs or sperms are discharged. One genital plate (the right anterior) is also the MADREPORITE and is finely perforated. In irregular echinoids (fig. 71i,j), the apical system does not enclose the periproct and it is small and compact; it may contain less than five genital plates, but always has five oculars.

CORONA. In Mesozoic and Cainozoic echinoids, with rare exceptions, the ambulacra and interambulacra each consist of paired columns of plates. In Palaeozoic echinoids, however, each may contain more than two columns.

62   A modern regular sea-urchin with long spines. Longest spines about
75 mm. *Astropyga magnifica* in 26 m of water off Florida Keys.
(Underwater photograph.)

63   A fossil echinoid with spines preserved *in situ*. *Cidaris clavigera*, Upper Chalk, Upper Cretaceous (×1·6).

The AMBULACRAL plates are small and each is pierced by one pair of pores, A PORE PAIR (fig. 71d), except in the gill-bearing, regular echinoids which have COMPOUND ambulacral plates (fig. 61b). These consist of two or more plates fused together and have two or more pairs of pores. The pores are round and close-set, except in the irregular echinoids which possess respiratory tube feet; in these, one or both pores of each pair may be elongated (fig. 67a). INTERAMBULACRAL plates (fig. 61b) are large and have no pores. Their surface is covered by many tubercles and granules to which, in life, movable spines are attached by muscles. The spines are rarely preserved, *in situ*, in fossil echinoids.

A TUBERCLE (fig. 61d, e) consists of a round knob, the MAMELON, protruding from a shallow mound, the BOSS. Tubercles occur on both ambulacral and interambulacral plates, but they are more numerous and larger on the latter. In regular echinoids they vary in size from large primary tubercles to small granules; but they are rather small, and close-set, in irregular echinoids.

A SPINE (fig. 61d) consists of a shaft with a socket at its proximal end which fits over the mamelon of a tubercle. Spines are appropriate in size to the tubercles with which they articulate. Thus, in regular echinoids, spines are of several sizes, whereas in irregular forms they tend to be more uniform in size and are close-set like a bristly fur. Spines are in part a protective device and in part are used for walking or burrowing. Thus, they vary widely in shape; they may be needle-like, rod-shaped, club-shaped or spatulate (figs. 62, 63).

The lower edge of the corona, to which the peristome membrane is attached, is referred to as the PERISTOME MARGIN. This may be an entire margin, i.e. a simple, circular opening, but, in some regular echinoids, it is notched for the passage of external gills (fig. 61g). In both regular and irregular echinoids which have jaws, the peristome margin projects inside the test as a series of processes, the PERIGNATHIC GIRDLE, which may form more or less complete arches over the radial water vessels (figs. 61h and 64).

The JAWS (Aristotle's lantern, figs. 61f and 65) consist of a framework of calcareous pieces, usually 40, of which five are sharp chisel-like TEETH protruding from the mouth. The teeth are moved out and in by a complicated system of muscles attached to the jaws and to the perignathic girdle.

Modifications of the test occurring in irregular echinoids

POSITION OF THE ANUS. In irregular echinoids, the anus may remain on the aboral side, either flush with the surface or in a groove (fig. 73a); or it may lie on the margin, or on the oral side, of the test (fig. 67a).

AMBULACRA AND PORE PAIRS. In most irregular echinoids the form of the pore pairs and of the ambulacra shows some modification, especially on the aboral side adjacent to the apical system where the tube feet have a respiratory function. Some forms may show only a slight elongation of the *outer* pore of each pore pair. In others, the two rows of pore pairs in each ambulacrum diverge and then converge, and, together, the ambulacra resemble a five-rayed flower; hence they are described as PETALS (or PETALOID) (fig. 67a). The pore pairs in the lower part of the test, beyond the petals, are not usually modified, but they may be few in number.

ORAL SURFACE. In the irregular echinoids, in which the mouth lies towards the anterior end, jaws and perignathic girdle are lacking. The

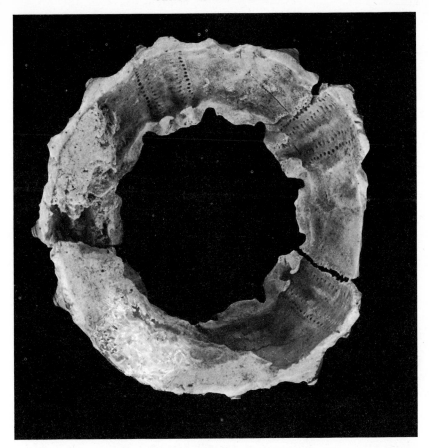

64   Peristome margin, with perignathic girdle and gill notches, viewed from inside the test of a regular echinoid. *Hemicidaris,* **Coral Rag, Corallian, Upper Jurassic, Wilts. (×3).**

The matrix adhering to the test in this fossil, and also the one in fig. 65, was removed (by Dr P. M. Kier) by means of a jet of compressed air containing an abrasive.

posterior interambulacrum is extended towards the mouth as a broad ridge, the PLASTRON (fig. 74e), and, at its forward end, may project as a lip, the LABRUM, on the lower side of the mouth (fig. 74e). The plastron is flanked by similarly extended posterior ambulacra.

FASCIOLES are relatively smooth bands on the test of certain irregular echinoids like *Micraster* (fig. 74e). The bands are covered by minute granules to which, in the living echinoid, very fine, ciliated spines were attached. Fascioles are named according to their location, e.g. a subanal fasciole lies below the anus.

65   Echinoid jaws with teeth. An oral view of '*Cidaris*', Coral Rag, Upper Jurassic (×1·5).

Test form and mode of life in some echinoids

Echinus (fig. 61) is a regular echinoid. The test is hemispherical, flattened on the oral side, and with a circular outline when viewed from the apex. The apical system is about one-fifth of the diameter of the test, and the plates form two rings, with the genital plates in contact with the periproct. The ambulacra are about half the width of the interambulacra, and the plates are compound, each with three pairs of pores arranged in three vertical rows. The primary tubercles form vertical series on both ambulacra and interambulacra, and there are many smaller tubercles and granules dispersed between them. There are gill notches on the peristome margin. Jaws and a perignathic girdle are present.

Pliocene to Recent.

66 *Echinus esculentis* clinging to rock; some of the tube feet can be seen extending beyond the spines. Diameter, about 90 mm. The small echinoid at the bottom right, holding a shell over its apical system, is *Psammechinus miliaris*.

(Photographed in shallow water, at low spring tide, on the west coast of Argyll.)

Echinus esculentus is a common British echinoid which lives mainly on a hard sea-floor in the sublittoral zone, and which ranges down to about 1000 m. The tube feet, which may extend beyond the spines, can grip a rock firmly so that the echinoid is not easily dislodged by waves. *Echinus* is partly carnivorous, eating small animals such as bryozoans and worms, but it also browses on seaweed. It moves slowly using its tube feet when on rock and its spines when on sand.

Clypeaster (fig. 67a, b) is an irregular echinoid with a flattened test of oval outline. The apical system, small and compact, consists of a perforated, star-shaped plate (representing both genital plates and madreporite) and five ocular plates. The ambulacra are petaloid. The periproct lies on the oral side near the posterior margin. Five radial grooves converge on the mouth which lies in the centre of the oral surface and contains massive jaws.

Upper Eocene to Recent.

Clypeaster lives today in tropical waters. It occurs in close-packed colonies; individuals are covered to a varying extent by a thin layer of sand or shell debris (fig. 68). It feeds on organic debris extracted from sediment and shovelled up with its teeth.

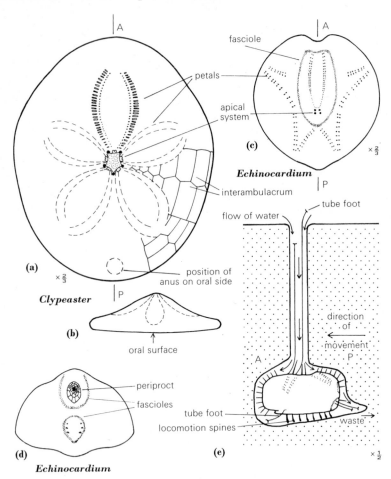

67   Mode of life of irregular echinoids.

a, b, *Clypeaster;* a, aboral view (note that the anus is on the oral side); b, side view. c–e, *Echinocardium;* c, aboral view; d, view of posterior end; e, diagram of echinoid in its burrow (only some of the tube feet are shown extended; the small arrows show the direction of movement of currents of water).

68 *Clypeaster* in its natural environment. *Clypeaster rosaceus* almost completely covered with sand and shells, including the empty test of an echinoid.

(Underwater photograph in 7 m of water off the Florida Keys, U.S.A.)

Echinocardium (fig.67c–e) is an irregular echinoid with a heart-shaped test. The anterior ambulacrum lies in a groove leading to the mouth, and has simple pores; the other ambulacra are subpetaloid. The periproct lies on the posterior margin. Fascioles are present within the petals and below the periproct (subanal). A plastron with relatively coarse tubercles is developed. The mouth lies near the anterior end and opens forwards. The spines are short and curved and are directed towards the posterior end; those on the plastron are paddle-shaped and are slightly larger than those on the rest of the test.

Oligocene to Recent.

Echinocardium cordatum is common near low tide mark on sandy beaches round Britain. It burrows in the sand to a depth of about 150 mm, keeping contact with the sea-water through a slender vertical pipe (fig.67e). To move the sand, the animal uses its spines and also certain tube feet, in particular those of the anterior ambulacrum. These tube feet can be extended to a considerable length and they construct the

69   Sand dollars in motion. Two specimens of *Encope michelini* in process of disappearing under a thin layer of sediment as they move forwards; time interval between the two exposures, about five minutes.
(Underwater photograph in 4 m of water off the Florida Keys, U.S.A.)

upper part of the vertical tube and stabilise the sand grains around it with a secretion of mucus. A current of water is drawn down the pipe by the ciliary action of the specialised spines on the fasciole within the petals. Oxygen is thereby brought to the respiratory tube feet of the latter. The animal feeds on organic debris within the sand, using spines and tube feet to convey particles to the mouth. There is a 'sanitary' pipe at the rear of the burrow and waste is flushed away along this by a current created by the cilia of the subanal fasciole. When the immediate supply of food is used up, the echinoid burrows forward, and makes a fresh pipe to the surface.

SUMMARY Echinoids may be assigned broadly to one of three groups, each following a different mode of life: (i) vagrant regular forms, with a more or less globular test, which use tube feet and spines for locomotion; (ii) irregular forms, with a more or less flattened test, which live, partly or completely, covered by a thin layer of sediment (fig. 69); (iii) irregular forms, with a heart-shaped test, which live in burrows in soft sediment. All forms are gregarious.

Additional common genera
Palaeozoic echinoids

The Palaeozoic echinoids were all regular forms. With two exceptions, the corona consisted of more than 20 columns of plates; the number varied between 2 and 20 in each ambulacrum, and 1 and 14 in each interambulacrum. In some forms the plates were thick with bevelled edges, while in others the plates were arranged in overlapping series, rather like fish scales, and the test appears to have been flexible. The ambulacral plates were simple, each having one pore pair. Jaws have been found in many genera, but there was no perignathic girdle. Entire tests are rare, but separated plates may be common in places, and are immediately distinguished as echinoid, since no other echinoderm has either pore pairs or tubercles.

One order, the cidaroids, may be referred to at this point, since it includes *Miocidaris* (Permian to Jurassic), the only echinoid known to have survived the Palaeozoic. *Miocidaris* is the earliest known genus to possess 20 columns of plates in the corona, and a perignathic girdle. It is, possibly, the stock from which the post-Palaeozoic echinoids were derived.

Archaeocidaris (fig. 71a) is a cidaroid with two columns of plates in each

70 A Carboniferous echinoid; *Melonechinus*, Mississippian (Lower
Carboniferous), Missouri, U.S.A. ($\times 0{\cdot}8$). Lateral view of a flattened test.

ambulacrum and four columns in each interambulacrum. The test was
flexible.

Lower Carboniferous.

Post-Palaeozoic echinoids

Post-Palaeozoic echinoids have, typically, a rigid test containing 20
columns of plates in the corona. They fall into two groups, the cidaroids,
whose affinities are with the Palaeozoic echinoids, and the euechinoids,
which appeared at the beginning of the Mesozoic. The cidaroids are
regular forms with simple ambulacral plates and without gills and gill
notches. They show little change in their main characters in the course
of their history. The euechinoids contain the vast majority of the post-
Palaeozoic echinoids. These show great diversity of form, including
regular forms with compound ambulacral plates and with gills and gill
notches, and also irregular forms.

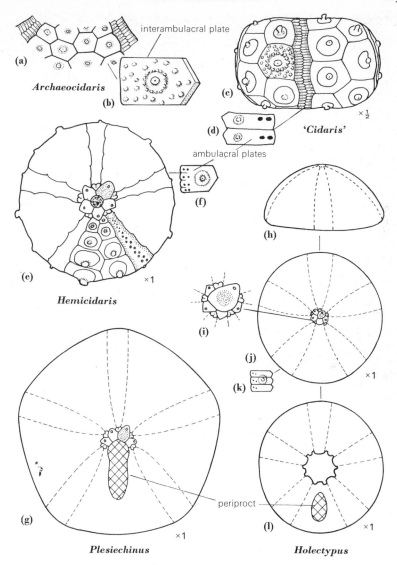

interambulacral plate

(a)

Archaeocidaris

(b)

(c)

(d)

×½

'Cidaris'

ambulacral plates

(f)

(h)

(e)

×1

Hemicidaris

(i)

(j)

×1

(k)

(g)

periproct

(l)

×1

×1

Plesiechinus

Holectypus

71 Regular and irregular echinoids.

CIDAROID

'*Cidaris*' (figs. 71c and 63), has a hemispherical test, flattened on oral and aboral surfaces. The apical system is large, about one-half the diameter of the test. The ambulacra are narrow and sinuous and have simple plates. The interambulacral plates each bear a large primary

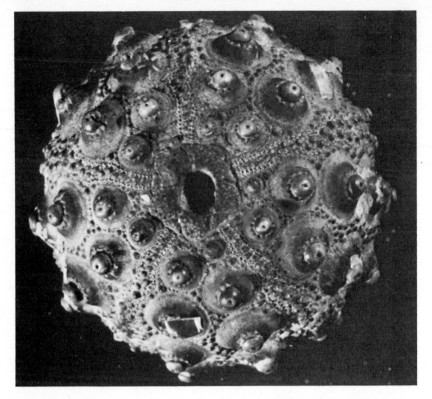

72   *Hemicidaris*, Coral Rag, Upper Jurassic, Wilts. (×3). Aboral view.

tubercle, surrounded by much smaller tubercles. The primary spines are very large and stout. There are jaws and a perignathic girdle, but no gill notches. *Cidaris* lives today in warm waters. It walks, using its large spines rather like stilts.

Jurassic to Recent.

EUECHINOIDS

(i) Regular form.

Hemicidaris (figs. 71e and 72) has a hemispherical test. The ambulacra are narrow and slightly sinuous. The plates on the aboral surface are simple, but towards the oral surface they are compound, each made of two to four fused plates. The interambulacral plates each bear a large primary and many smaller tubercles with long slender spines. The peristome margin has gill notches.

Middle Jurassic to Upper Cretaceous.

(ii) Irregular forms with the mouth placed centrally, and jaws.

Plesiechinus (fig. 71g) has a depressed test with a roughly pentagonal outline. The anus is in contact with the posterior margin of the apical system. The ambulacra have simple plates with round pores.
Lower to Middle Jurassic.

Holectypus (fig. 71i–l) is similar to *Plesiechinus*, but the anus is on the oral surface.
Lower Jurassic to Upper Cretaceous.

(iii) Irregular forms with the mouth lying near the anterior margin.

Nucleolites (fig. 73a) has a test of oval outline. The anus lies in a groove on the aboral side. The ambulacra are subpetaloid, and the outer pore of each pore pair in the petals is elongated.
Middle Jurassic to Upper Cretaceous.

Clypeus (fig. 73d–f) is similar to *Nucleolites*, but the test is more depressed

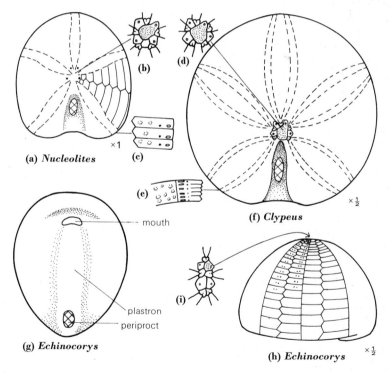

(a) *Nucleolites*   (b)   (c)   ×1

(d)

(e)

(f) *Clypeus*   ×½

mouth

plastron
periproct

(g) *Echinocorys*

(i)

(h) *Echinocorys*   ×½

73   Irregular echinoids.

and almost discoidal in shape. The ambulacra, too, are more distinctly petaloid, and the outer pore of each pore pair is greatly elongated.

Middle and Upper Jurassic.

Echinocorys (fig. 73g–i) has a high test which is strongly convex on the aboral side, and has an oval outline. The apical system is elongated. The three anterior ambulacra are quite widely separated from the two posterior ambulacra. The pores are round and their number is reduced towards the oral surface.

Upper Cretaceous.

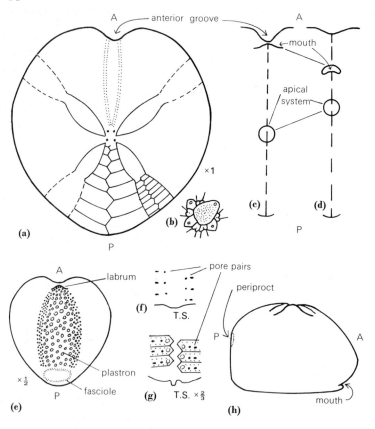

Micraster

74   Changes in the test in *Micraster*.

a, aboral view of a high zonal form. b, apical system. c, d, a comparison of the positions of the apical system and the mouth, and also the depth of the anterior groove, in a high zonal *Micraster* (c), and a low zonal *Micraster* (d). e, oral surface of a high zonal form. f, g, pore pairs and cross-section of an ambulacrum of a low zonal form (f), and a high zonal form (g). h, side view of a high zonal form.

Holaster is related to *Echinocorys*, from which it differs in having a heart-shaped test, with the anterior ambulacrum lying in a groove which leads down to the mouth.

Cretaceous.

Micraster (figs. 74 to 77) has a heart-shaped test with a small, compact apical system. The anterior ambulacrum is in a deep groove leading down to the mouth; the other ambulacra are subpetaloid. The three anterior ambulacra form a group separate from the two posterior ambulacra. The anus lies on the upper part of the posterior margin and, below it, is a small fasciole. The plastron is broad, and the labrum is prominent.

Upper Cretaceous.

A series of changes occurred in *Micraster*, during the Upper Cretaceous, which provides a classic example of evolution and which is also of great value stratigraphically. The echinoid occurs throughout several hundred feet of the Middle and Upper Chalk. Specimens from the lowest and highest beds differ considerably and are referred to different species. Specimens from the intervening beds, however, are transitional between these two species and show clearly that the differences between them are the result of gradual changes which occurred in a number of characters.

Some of the main changes may be summarised thus: (i) the test became broader and higher; (ii) the anterior groove, leading to the mouth, deepened, thus emphasising the heart-shaped outline of the test; (iii) the petals elongated; and the surface of the plates between the pore pairs, which was initially smooth, became granular; (iv) the mouth moved further forwards, and a projecting labrum developed; (v) the tubercles on and around the plastron coarsened; (vi) the fasciole below the anus became broader. Some of these changes are shown in figs. 74 to 77.

There is, of course, some variation in the development of these (and other) characters in a collection of *Micrasters* from any particular level in the Chalk, and one specimen may show a combination of characters some of which are more, and others less, advanced.

The comparatively uniform character of the Chalk probably signifies that the environment, in which *Micraster* lived, did not vary greatly; and so it is reasonable to conclude that the evolutionary changes in this echinoid were connected with its adaptation to a particular mode of life. A recent study of the living habits of modern 'heart' echinoids including

75   *Micraster;* aboral view of a high zonal form (×2).

Echinocardium (p. 115) has led to the suggestion that the evolution of *Micraster* represents a progressive adaption to burrowing more deeply in the Chalk sediment. Thus, for example: (i) the increased number of large tubercles on the plastron suggests more and stronger digging spines; (ii) the granules on the petals presumably carried more cilia, to improve the flow of water through the burrow; (iii) the longer petals carried more respiratory tube feet and so improved extraction of oxygen from the water; (iv) the forwards shift of the mouth and possibly, also, the deepening of the anterior groove enabled more effective food-

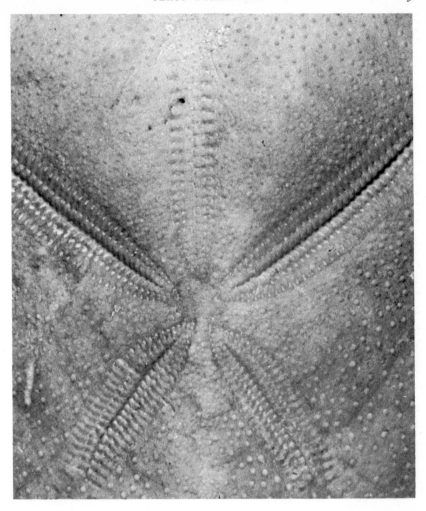

76 *Micraster;* petals of a high zonal form ($\times 4$).

gathering; (v) the broader fasciole implies extra cilia to improve the removal of waste.

Geological history of the echinoids

SUMMARY The first echinoid occurs in the Upper Ordovician, but further examples are rare until the Carboniferous. Numbers were reduced drastically at the end of the Palaeozoic, but they revived in the Mesozoic. They remained common in the Tertiary and are still a flourishing group.

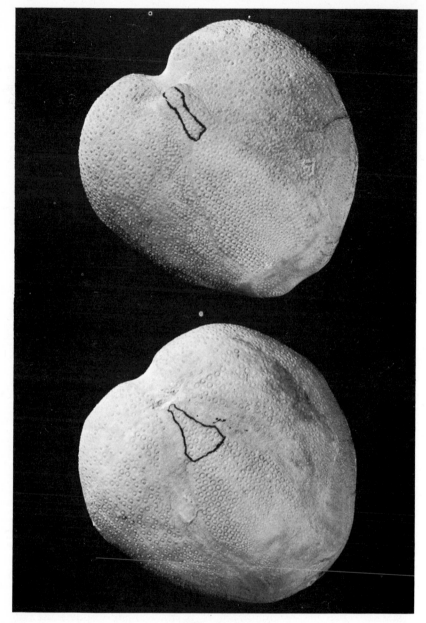

77   *Micraster;* oral view of a high zonal form (above) and a low zonal form
(below) ($\times 1.5$).

In these photographs, compare the depth of the notch formed in the outline by the anterior
groove; the development of the labrum; and the nature of the tubercles on and around the
plastron. The ink lines are drawn round the plate forming the labrum.

Echinoids were inconspicuous in the faunas of the Palaeozoic seas. The first example (*Bothriocidaris*, Upper Ordovician) is unique in having only 15 columns of coronal plates. Typically, however, the archaic echinoids had more than 20 columns of plates (p. 117). Most genera occur in the Carboniferous, and are usually found in limestones. *Archaeocidaris* (fig. 71a) is an example. There was a wholesale extinction of echinoids at the end of the Palaeozoic and only one genus, *Miocidaris* (Permian to Jurassic), a cidaroid, survived the Permian. Cidaroids like '*Cidaris*' (fig. 71d) are common, at times, in the Jurassic and Cretaceous rocks.

Euechinoids, derived from 'cidaroid' stock, appeared in the Trias. The first were regular, gill-bearing forms with compound ambulacral plates. They are common in the Mesozoic, but show little modification of the basic plan seen, for example, in *Hemicidaris* (Jurassic to Cretaceous, fig. 72). The first irregular echinoids are found in the Lias and, apart from the eccentric position of the anus, are similar to contemporary regular euechinoids, with which they share a common ancestry. Irregular echinoids appeared in increasing numbers during the later Jurassic and Cretaceous.

The Mesozoic echinoids are rarely found in clayey rocks, but may be very abundant in calcareous rocks. They occur, for instance, in the Clypeus Grit (Inferior Oolite), where *Clypeus* (fig. 73f) occurs in sufficient numbers to give the rock its name ('grit', here, refers to shell fragments in the limestone). The Chalk, too, is comparatively rich in echinoids, and they provide several zonal indices. *Micraster* (fig. 75) and *Echinocorys* (fig. 73g) are common forms.

Echinoids, especially irregular forms, are numerous in Cainozoic rocks in some regions of the world. In England, however, they are rare, since the Tertiary rocks here are mainly clays and non-marine deposits.

TECHNICAL TERMS

ABORAL the upper side of the test opposite the mouth (fig. 61a).

AMBITUS the circumference of the test.

AMBULACRA five, radially arranged columns of pore-bearing plates, extending between the mouth and the apical system, and overlying the radial water vessels (fig. 61a).

APICAL SYSTEM small plates (oculars and genitals) terminating the ambulacra and interambulacra at the apex of the test (fig. 61c).

ARISTOTLE'S LANTERN see jaws.

COMPOUND PLATE an ambulacral plate composed of two or more plates united by a primary tubercle, and bearing two or more pore pairs (fig. 61b).

CORONA the main part of the test, composed of five ambulacra and five interambulacra.

FASCIOLE a narrow band of ciliated spines, which articulate with very minute tubercles, and so appears on the naked test as a relatively smooth band (fig. 67c).

GENITAL PLATE a plate lying in the apical system, at the top of an interambulacrum, and pierced by a pore for the discharge of eggs or sperms (fig. 61c).

GILL SLITS notches in the peristome margin for the passage of external gills in regular echinoids, like *Echinus* (fig. 61g).

INTERAMBULACRA five columns of plates, lacking pores, which lie interradially, between the ambulacra (fig. 61a).

JAWS a complex series of calcareous pieces holding five sharp teeth, possessed by regular, and some irregular, echinoids (figs. 61f and 65).

LABRUM a lip formed by the end plate of the posterior interambulacrum, which projects below the mouth in forms like *Micraster* (fig. 74e).

MADREPORITE a perforated genital plate in the apical system (right anterior side), through which water enters the water vascular system (fig. 61c).

OCULAR PLATES five plates lying (in the apical system) one at the top of each ambulacrum (fig. 61c). Each is pierced by the terminal tentacle of the radial water vessel which was once thought to be light-sensitive; this explains the term 'ocular'.

ORAL SURFACE the lower side of the test, on which the mouth lies.

PEDICELLARIAE tiny pincer-like grasping organs, borne on granules, and used in defence and in cleansing the test.

PERIGNATHIC GIRDLE an extension, inside the test, from the peristome margin, to which the jaw muscles are attached (fig. 61h).

PERIPROCT a plated membrane which surrounds the anus (fig. 61c).

PERISTOME a plated membrane surrounding the mouth (fig. 61g).

PETAL the part of an ambulacrum, shaped like a petal, lying adjacent to the apical system in irregular echinoids with respiratory tube feet (fig. 67a).

PLASTRON a broad, inflated extension of the posterior interambulacrum occurring on the oral side, towards the mouth, in irregular echinoids like *Micraster* (fig. 74e).

PORE PAIR two close-set openings in an ambulacral plate, through which a single tube foot passes (fig. 71d).

SPINE a calcareous rod, articulating with a tubercle, and used in locomotion and defence; the largest are primary spines (fig. 61d).

TEST the entire echinoid skeleton, comprising periproct, apical system, corona and peristome.

TUBE FEET slender, extensible tentacles connected with the water vascular system and found in the ambulacra of all echinoderms. They function mainly in locomotion, feeding or respiration (fig. 61i).

TUBERCLES knobs occurring on most plates of the test, to which movable spines are attached (fig. 61e).

WATER VASCULAR SYSTEM the hydraulic system of tubes, through which water is circulated to the tube feet (fig. 61j).

10

Echinodermata, class Crinoidea and minor echinoderms

CLASS CRINOIDEA

Crinoids have a compact, cup-shaped body with the mouth in the centre of the upper, ORAL, surface and the anal opening near it. Five simple, or branched, flexible ARMS bearing ambulacral grooves, surround the mouth. The body is attached to the sea-floor, at least for a time, by a stem growing down from the under, ABORAL, side.

The endoskeleton consists of many calcareous plates, arranged with pentamerous symmetry. Those enclosing the body form the CALYX; further plates form the axes of the stem and arms.

Crinoids are gregarious animals which range in distribution from tropical seas to arctic waters. Living forms include sea-lilies (sessile forms with stems), found in deep water, and feather stars (pelagic stemless forms) which live in shallow, clear, coastal waters. Feather stars are more numerous and widespread today than sea-lilies, though the reverse is true of fossil crinoids.

Morphology

In living crinoids, the bulk of the soft body is contained in the calyx. This consists of two parts, the DORSAL CUP, covered on the oral side by a plated membrane, the TEGMEN. The arms articulate freely with the calyx. In many genera they branch, and each branch may bear two rows of small branchlets, PINNULES. Together they form an open 'funnel' around the mouth and each bears a ciliated food-groove, (AMBULACRUM), on its oral side, lined with tube feet which trap food (falling plankton) in mucus. The food is swept by ciliary action down the food-grooves to the mouth. The stem may be quite short, or many feet in length. It may be anchored in sediment by 'rootlets' diverging from its base; or it may have prehensile branches (cirri) with which the crinoid clings to seaweed.

SKELETON. The plates of the crinoid skeleton are disposed in three main regions, the CALYX, ARMS and STEM (fig. 78b). The calyx may be a relatively flexible structure, but in many fossil genera the plates were united to form a rigid cup or box. The individual plates are usually hexagonal or pentagonal in shape, and are arranged symmetrically in circlets.

The DORSAL CUP (on the aboral side of the calyx) consists of two circlets of five plates, the BASALS below, and the RADIALS above. In some forms an extra ring of five plates, the INFRABASALS, is intercalated between the basals and the stem (fig. 78b). In certain crinoids the dorsal cup contains additional plates; for instance, there may be INTERRADIAL plates intercalated between the radials; or in, some cases, the lowest arm plates, BRACHIALS, may be incorporated in the cup (fig. 79a).

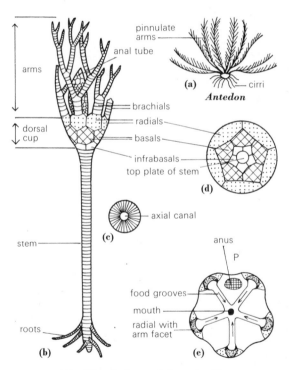

78 Morphology of the crinoids.

a, a free-swimming crinoid. b, an attached crinoid showing the disposition of the main parts of the body; the arms are incomplete. c, the articular surface of a stem plate (columnal). d, aboral view of the dorsal cup. e, oral view of the calyx to show the food grooves converging on the mouth.

The TEGMEN (on the oral side of the calyx) may be a membrane studded with discrete plates, or a rigid structure doming over the mouth and adjacent food-grooves. The ANAL OPENING lies in the posterior inter-radius (fig. 78e) between two food-grooves and is either on the surface of the tegmen or at the tip of a tube, the ANAL TUBE (fig. 78b).

The arm plates (BRACHIALS) articulate freely. They are cylindrical in shape, with a V-shaped incision, on the oral side, to accommodate the food-groove. The stem plates (COLUMNALS) also articulate freely. They are discoid plates of circular (fig. 78c) or star-shaped outline (fig. 79d), and each has a central hole through which an extension of the soft parts passes.

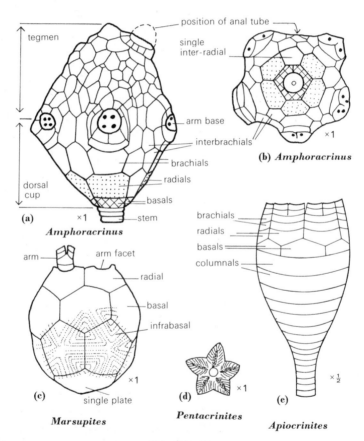

79   Crinoids.

b, aboral view of dorsal cup. In (c) the surface marking is drawn in on two plates only.
d, articular surface of a columnal.

Skeletal form and mode of life of some crinoids

Antedon (fig.78a) is a stemless form with a tiny calyx and pinnulate arms. It has a cluster of cirri at the base of the calyx.

Eocene to Recent.

Antedon bifida is the only crinoid found in British waters. It lives in the littoral and sublittoral zone, clinging to stones with its cirri, or swimming by gently waving its arms.

Apiocrinites (fig.79e) has a long stem which widens and gradually merges into the rigid, pear-shaped calyx. This is made up of the topmost stem plates, which are much enlarged and joined to the basals, and radials with some brachial plates. There are ten arms, not usually preserved in fossils.

Jurassic to Recent.

Apiocrinites lives in abyssal depths attached to the sea-floor by its stem.

SUMMARY Living crinoids fall into two main groups illustrated by the two examples described. They are either free-swimming, stemless forms, like *Antedon*, which live mainly in relatively shallow coastal waters, or stemmed forms which are often attached to the sea-floor in deeper waters, ranging down to about 5000 m. Fossil crinoids were also either stemmed or free-swimming, but geological evidence suggests that the habitat, especially in the case of stemmed forms, may have differed to some extent from that of living crinoids. An example is *Apiocrinites*, which today is an abyssal form, but which occurs commonly as a fossil in the Bradford Clay, a thin bed in the relatively shallow-water Great Oolite Series. Typically, too, the Palaeozoic crinoids, which were stemmed forms, are associated with a variety of benthonic organisms in shallow-water sediments, e.g. *Amphoracrinus* occurring in Lower Carboniferous 'reef' limestones.

Additional common genera

Palaeozoic crinoids

Palaeozoic crinoids were stemmed forms. Most had a rigid calyx and they show modifications of the arms, including the development of pinnules, which increased the effective area of food-gathering.

Amphoracrinus (fig.79a) has a rigid, ovoid calyx. The dorsal cup consists of three basals, five radials and five brachials. There is a single

80 *Pentacrinites*, Lower Lias, Jurassic (×2).

inter-radial plate. A small number of interbrachials are intercalated between the brachials. The tegmen forms a high domed roof of polygonal plates. There is a short anal tube. The arms (seldom preserved) branch several times.

Lower Carboniferous.

81 A Silurian crinoid. *Gissocrinus*, Wenlock Limestone (×2).

82   A Cretaceous (stemless) crinoid. *Uintacrinus*, Upper Cretaceous (×1).

Mesozoic and Cainozoic crinoids

Post-Palaeozoic crinoids all belong to the one subclass; they are characterised by a small calyx with a flexible tegmen, on which the mouth and
food-grooves are exposed. The arms articulate with the radials. Some,
like *Pentacrinites* and *Apiocrinites*, have stems; others, like *Marsupites*,
lack stems.

Pentacrinites (fig. 80) has a small calyx composed of infrabasals, basals
and radials; the arms are very long, much branched and bear pinnules.
The stem is long and has cirri at intervals; the columnals are star-
shaped (fig. 79d). *Pentacrinites* occurs in the Lower Jurassic in England.
It survives today in moderately deep water. It breaks free from its
mooring when adult, and swims, trailing its long stem.
   Trias to Recent.

Marsupites (fig. 79c) has a large, stemless, globular calyx. The individual
plates are large and are pentagonal in shape, except the basals, which are
hexagonal; they each bear a ridged pattern. The base of the calyx is a
single plate, around which are infrabasals followed by basals and
radials. The radials have notched facets for the arms. The arms, which
are short and fork repeatedly, are rarely preserved.
   Cretaceous.

Geological history of the crinoids

SUMMARY The earliest record of crinoids is from the Arenig, and they were the commonest echinoderms during the Palaeozoic. Their numbers went down for a time in the Permian, but revived in the Mesozoic, although they did not again reach their Palaeozoic peak.

Crinoids first became abundant as fossils in Silurian rocks, and they remained numerous in the Devonian and Carboniferous. They occur, most typically, in calcareous rocks, e.g. the Wenlock Limestone (fig. 81) or Carboniferous reef limestones. An entire calyx is rare but separate plates, especially columnals, may be abundant and may be the main constituent of a rock which is then called a 'crinoidal' limestone. *Amphoracrinus* is one of the common Carboniferous crinoids (fig. 79a). An entire calyx may occasionally be collected from some of the reef knolls in the North of England.

None of the Palaeozoic groups of crinoids survived into the Mesozoic, but, while the Mesozoic crinoids must have descended from the earlier forms, the relationship is not clear. The earlier Mesozoic crinoids were

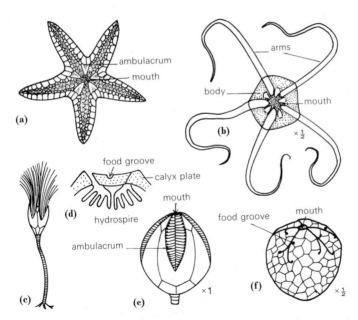

83   Minor echinoderms.

a, starfish, oral view. b, ophiuroid, oral view. c–e, blastoid; c, restoration of a complete specimen; d. section across an ambulacrum; e, lateral view of the calyx. f, cystoid; lateral view of
the theca showing the irregular arrangement of plates.

84   A blastoid. *Pentremites*, Mississippian (Lower Carboniferous) ($\times 2.7$).
Left, lateral view of calyx. Right, oral view.

mainly stemmed forms, e.g. *Pentacrinites* (fig. 80), but later in the
Cretaceous, stemless forms, such as *Marsupites* (fig. 79c) and *Uintacrinus*
(fig. 82) became the more abundant. They remain the dominant crinoids
today.

MINOR ECHINODERMS

Cystoids

Cystoids (fig. 83f) are distinguished by a unique system of pores and
canals in some, or all, of the plates that make up the rigid test or THECA.
The theca, of ovoid or globular shape, was attached, directly or by a
stem, to the sea-floor. It consists of many polygonal plates arranged
haphazardly or regularly. Pentameral symmetry is not highly developed,
especially in earlier forms. The mouth and anus are on the upper (oral)
surface. AMBULACRAL GROOVES (along which food was carried to the
mouth) radiate from the mouth over the surface of the theca and may,
in some forms, extend along BRACHIOLES, erect food-gathering 'arms'.

Cystoids are rare fossils. They range from the Lower Ordovician to the Middle Devonian, reaching an acme in the Silurian. They occur usually in calcareous rocks, for instance in the Wenlock Limestone, and are often associated with reef formations.

Blastoids

Blastoids (figs. 83c–e and 84) have a bud-shaped test, the CALYX, consisting of three circlets, each of five plates, arranged with pentameral symmetry. The calyx was attached aborally either by a stem, or directly, to the sea-floor. The mouth was in the centre of the upper surface, surrounded by five pores, one of which was the anal opening (fig. 84). Five food-grooves radiated from the mouth, over the surface of the calyx and up the delicate arms. Blastoids are distinguished by paired, internal folds of skeletal tissue, the HYDROSPIRES, which underlie the food-grooves and which are presumed to have aided in respiration.

Blastoids ranged from the Silurian to the Lower Permian, and were at their acme in the Carboniferous. They are rare as fossils; occasionally, however, they may be found in considerable numbers in calcareous rocks of shallow-water origin, often associated with 'reef' limestones, e.g. in the Lower Carboniferous.

85   A starfish (*Oreaster reticulatus*) preying on an echinoid (*Meoma ventricosa*).
(Underwater photograph in 3 m of water off the Florida Keys, U.S.A.)

86 A Silurian ophiuroid. *Lapworthura miltoni*, Lower Ludlow,
Leintwardine, Hereford (×1·5).

87 A Jurassic ophiuroid. *Ophiurella speciosa*, Lithographic Limestone,
Upper Jurassic, Solenhofen, Bavaria (×0·7).

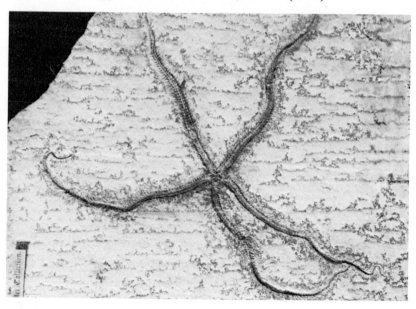

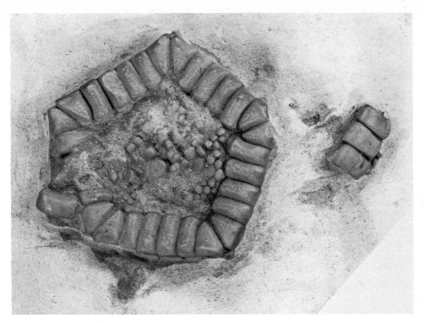

88 A Cretaceous Starfish. *Metopaster*, **Upper Cretaceous (×1·7).**
Aboral view of a pentagonal form with very short arms, showing the arrangement of large
plates around the margin, and smaller plates covering the remainder of the surface.

Stelleroids

The stelleroids, including starfish (fig. 83a) and ophiuroids (figs. 83b, 87), are star-shaped animals with a depressed, disc-like body, from which five or more arms radiate. The mouth is in the centre of the lower (oral) surface, and ambulacra, with tube feet, radiate from it along the arms. The ambulacra are open grooves in starfish, but lie within the arms in ophiuroids. The body is flexible and can, to some extent, change shape. The skeleton consists of discrete, internal plates. Modern stelleroids are usually vagrant, and live on sandy or rocky sea-floors (see fig. 85), but some members burrow in sediment.

Stelleroids range from the Tremadoc to the present day. While, in general, they are rare fossils, in some horizons (referred to as 'starfish' beds) they are relatively numerous, e.g. in the Upper Ordovician, Girvan and in the Lower Ludlow in Herefordshire (fig. 86). Starfish remains are also relatively common in the Chalk (fig. 88). Their fossils occur in muddy, sandy and calcareous sediments.

TECHNICAL TERMS (CRINOIDS)

ARM a jointed appendage bearing an ambulacral groove on its oral side and supported by calcareous ossicles, brachials (fig. 78b).

ANAL TUBE a tube, with the anal opening at its end, which projects from the oral surface (fig. 78b).

CALYX the structure, consisting of dorsal cup and tegmen, which contains the bulk of the crinoid body (fig. 79a).

CIRRI small prehensile branches from the stem (fig. 78a).

DORSAL CUP the aboral, cup-shaped part of the calyx, consisting of circlets of plates, including radials above and basals below. In some forms, infrabasals are intercalated between the stem and the basals (fig. 78b).

STEM the flexible stalk, supported by a series of columnal plates, by which the crinoid is usually fixed to the sea-floor (fig. 78b).

TEGMEN a membrane, with or without plates, which covers the oral side of the body (fig. 79a).

11

Brachiopoda

The Brachiopoda are sessile marine animals which secrete an external shell consisting of two dissimilar but equilateral valves. The plane of symmetry bisects the valves in an anterior-posterior direction. The number of living species is small and the group is only a minor one at the present day. Fossil brachiopods, however, were abundant and varied in Palaeozoic and Mesozoic rocks.

The shell may be calcareous or chitinous, and is secreted by the mantle which consists of two extensions of the body wall, one from the dorsal side and one from the ventral side of the body. The valves are thus DORSAL and VENTRAL in position. The brachiopod is attached to the sea-floor at its posterior end, usually by a fleshy stalk, the PEDICLE. The body lies in the posterior part of the shell, and from it the mantle lobes extend forwards to enclose a space, the MANTLE CAVITY, much of which is occupied by a food-collecting mechanism, the LOPHO-PHORE. There is a well-developed coelom. The nervous and circulatory systems are not highly organised and definite sensory organs are lacking. The sexes are typically separate; gametes are shed into the sea, or in some cases eggs may be brooded in the mantle cavity. The larvae are free-swimming for a short time before they settle down on to the sea-floor and metamorphose.

Brachiopods are distributed widely throughout the world in cool and temperate waters; they are commonest in eastern waters, e.g. around Japan, South Australia and New Zealand. Most live in the neritic zone, but a few range into deeper water down to about 5000 m.

Brachiopod shells range in size from a few millimetres to as much as 370 mm in width by 250 mm in length in the case of a Carboniferous form called *Gigantoproductus giganteus*. Most measure between about 20 mm and 70 mm.

Morphology

Brachiopods have been separated into two classes, the ARTICULATA and the INARTICULATA but there are enough similarities in their general morphology to consider them together.

The brachiopod shell encloses the body except for the pedicle (fig. 89d). The valve on the VENTRAL side of the body is known as the PEDICLE valve, since the pedicle commonly emerges through it, while the valve on the DORSAL side is the BRACHIAL valve, and takes its name from the brachia, arm-like projections, which make up the lophophore. Commonly the pedicle valve is the larger, projecting at its posterior end beyond the brachial valve. The pedicle emerges from the shell at its POSTERIOR margin and the opposite margin is ANTERIOR. The valves open slightly along the anterior margin during feeding (fig. 90e), but remain in contact along the posterior margin by means of a HINGE in the Articulata and by a system of muscles in the Inarticulata. SOFT BODY. The body and mantle line the shell and in some cases the soft tissue extends by minute tubules into the shell wall. The main part of the body is small, and much of the mantle cavity is taken up by the lophophore (fig. 89d). This may be a lobed disc or two coiled or folded arms called BRACHIA each of which has a groove leading back to the mouth and fringed with ciliated tentacles. These maintain currents of water along three paths, a median outgoing flow, and an incurrent flow on either side. Minute organisms, frequently diatoms, are filtered from the incurrent water and passed along the lophophore grooves to the mouth and thence to the digestive tract. The intestine ends blindly in living articulate brachiopods, but opens in an anus in inarticulates.

Most brachiopods are attached by a PEDICLE (fig. 89d) which typically is a stout fleshy stalk attached to the pedicle valve by muscles. Its distal end is fixed to a rock or shell, or may diverge into rootlets to secure a hold in soft sediment. In some forms the pedicle may be absent, in which case the shell is usually cemented to a firm surface (fig. 91d). Some extinct forms appear to have been anchored in soft sediment by spines (fig. 92i).

The opening and closing of the valves is controlled by a system of muscles which are attached to the inner surface of the valves towards the posterior end where they may leave MUSCLE SCARS. The muscle system is simplest in the articulate brachiopods (fig. 89c), consisting commonly of a pair of ADDUCTOR muscles, which run across the shell cavity from the interior of the pedicle valve to the interior of the brachial valve, and of two pairs of DIDUCTOR muscles which run obliquely from the pedicle valve to a projection, the CARDINAL PROCESS, from the hinge line of the brachial valve. Both sets of muscles work by contracting. The hinge line acts as a fulcrum, and the cardinal process as a lever so that, as the diductor muscles contract, they pull down the cardinal process and the valves open. As the diductor muscles relax, the adductor muscles contract and pull the valves together.

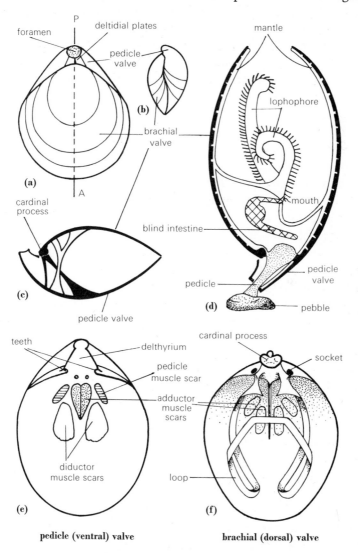

89. Morphology of an articulate brachiopod.

a, dorsal view (A, anterior; P, posterior; line A–P, plane of bilateral symmetry). b, side view. c, section to show the disposition of the adductor muscles (unshaded) and the diductor muscles (black). d, simplified section to show the general relationship of the soft part to the shell. e, interior of the pedicle valve. f, interior of the brachial valve.

There are additional muscles in inarticulate brachiopods which control lateral movement of the valves relative to one another.

SHELL. In most INARTICULATE brachiopods the shell has a horny appearance and is composed of alternate layers of chitin and calcium phosphate, but in a few forms it is calcareous. In ARTICULATE brachiopods the shell is calcareous. A thin chitinous layer (periostracum) overlies the shelly part which is made up of an inner and outer layer of calcite fibres. The inner layer thickens with growth. In some articulate groups, the shell is punctate, the inner surface being perforated by fine tubules (p. 144), while in other groups the shell lacks these tubules and is impunctate.

The shape of the shell may show some correlation with the arrangement of the lophophore and the feeding currents, and markings on the inner surface may provide information about the disposition of the muscles, the lophophore and canals in the mantle in a number of extinct forms. In inarticulate brachiopods the shell is approximately oval or circular in outline with gently convex valves (fig.91a,b). In the articulate forms the shell may be ovate (fig.89a), tapering slightly at the posterior end and with a short curved hinge line, or it may be semicircular in outline with a straight wide hinge line (fig.91h). The pedicle valve is typically larger than the brachial valve. Both valves may be convex, or one may be convex and the other flat or concave. The shell may be folded along its midline so that a ridge or FOLD is formed at the anterior margin of one valve (fig.94b) with a corresponding depression or SULCUS on the other. The fold and sulcus serve to keep separate the incoming and outgoing currents of water.

The brachiopod shell grows by increments to the margin which typically are greater along the anterior and lateral margins and which form concentric growth lines on the outer surface. Thus the initial shell remains at or near the posterior margin and may form the tip of a pointed BEAK. The curved convex area around the beak, which is generally a more prominent feature in the pedicle valve, is the UMBO and in many forms a curved or flat INTERAREA is interposed between it and the hinge line (fig.93a). The surface of the shell may be smooth, or may be marked by concentric or radial lines or ribs and tubercles or spines may be developed. This surface ornament is useful in distinguishing species.

The opening through which the PEDICLE emerges in inarticulate brachiopods is usually a gape, but may be a groove or a slit in the pedicle valve. In the articulate brachiopods the pedicle opening, the

DELTHYRIUM, is a triangular gap in the posterior margin of the pedicle valve (fig. 89e), and commonly is constricted by a pair of DELTIDIAL PLATES (fig. 89a), or by a single plate, the DELTIDIUM leaving a circular hole, the FORAMEN for the passage of the pedicle (fig. 89a).

The hinge apparatus consists of TWO TEETH in the pedicle valve (fig. 89e) which fit into TWO SOCKETS in the brachial valve (fig. 89f). The teeth are short projections from the hinge line, one on each side of the delthyrium and they may be supported in some genera by DENTAL PLATES projecting from the floor of the pedicle valve. The sockets lie, one on each side of a small projection, the CARDINAL PROCESS, to which the diductor (or opening) muscles are attached (fig. 89f).

In most articulate brachiopods there are distinct SCARS left on the floor of the valves by muscles. The degree to which they are defined and their relative positions, however, may vary in different genera. In the

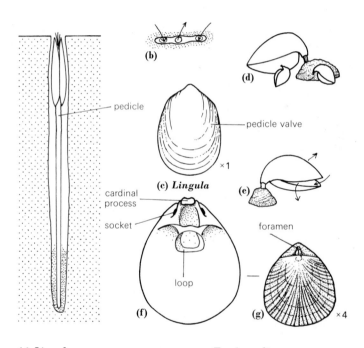

(a) *Lingula*

Terebratulina

90 Morphology and mode of life of brachiopods.

a–c, an inarticulate brachiopod, *Lingula;* a, in feeding position at the mouth of its burrow; b, slit-like opening of the burrow (the arrows indicate the incurrent and excurrent flow of water); c, a fossil *Lingula.* d–g, an articulate brachiopod, *Terebratulina;* d, attached to a stone and to other shells; e, in feeding position with the shell open (arrows as for b); f, interior of the brachial valve; g, brachial valve exterior.

pedicle valve, the pedicle and diductor muscle scars are grouped round two close-set adductor muscle scars (fig. 89e). In the brachial valve, four adductor muscle scars are grouped on the floor of the valve, and the diductor muscle scars are on the cardinal process (fig. 89f). In a small number of brachiopods, the muscles are attached to a trough-shaped structure, the SPONDYLIUM, in the pedicle valve (fig. 91j, k). This consists of two enlarged dental plates which converge, and may unite on the floor of the valve to form a trough which may be supported on a septum. There are additional muscle scars in the inarticulate brachiopods, but they are more complicated and often may be indistinct.

In some groups of articulate brachiopods the lophophore is supported by a calcareous framework, the BRACHIDIUM. This is a feature of systematic importance. It is attached to the hinge line of the brachial valve just below the sockets and may consist of either: (i) two small projecting prongs, CRURA (fig. 91f); or (ii) two long spirally coiled calcareous ribbons, SPIRALIA (fig. 93d); or (iii) two short ribbons united to form a LOOP (fig. 90f); or (iv) two longer ribbons united to form a REFLEXED LOOP (fig. 89f).

Shell form and mode of life of two brachiopods

Lingula (fig. 90a–c) is an inarticulate brachiopod with a translucent horny shell. It is spatulate in outline, tapering slightly towards the posterior end. The valves are almost equal in size and the pedicle opening is shared by both valves. There are regular growth lines on the outer surface of the shell. *Lingula* is a long-ranged genus, but individual species have usually a short range.

Silurian to Recent.

Lingula is found today mainly in Japanese coastal waters where it burrows in sand or mud, often in mudflats exposed at low tide. It differs in mode of life from a typical brachiopod in several respects. It lives, anchored by a long worm-like pedicle, in a burrow which may be 300 mm deep. It feeds, anterior end uppermost, at the slit-like opening of its burrow. At low tide, or when otherwise disturbed, it withdraws down its burrow by contracting its pedicle. Unlike other brachiopods, which typically live under fully marine conditions, *Lingula* can survive for a short period in tidal water made brackish by river flood-water. As a fossil, *Lingula* occurs commonly without other brachiopods in shaley beds.

Terebratulina (fig. 90d–g) is an articulate brachiopod. It has a biconvex shell of more or less oval outline, tapering towards the beak. The hinge

line is short and gently curved. The foramen is large and below it the delthyrium is closed by small deltidial plates. Fine ribs radiate from the beaks. The brachidium is a short ring-like loop.
Jurassic to Recent.

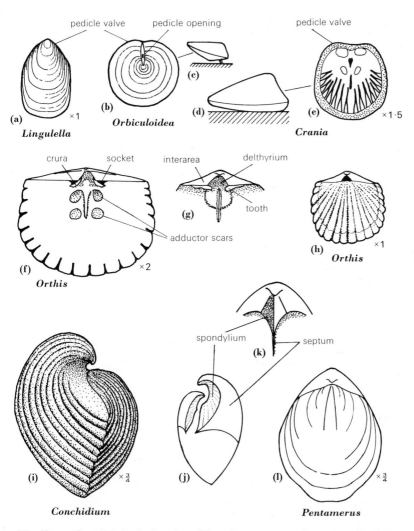

91 **Examples of inarticulate brachiopods, pentamerids and an orthid.**
a–e, inarticulates; a, b, pedicle valves; c, d, side view of shells *in situ*; e, interior of the pedicle valve. f–h, an orthid; f, interior of the posterior end of the brachial valve and g, of the pedicle valve; h, brachial valve. i–l, pentamerids; i, side view; j, almost median longitudinal section showing part of the spondylium supported by a septum; l, brachial valve showing the trace of the two septa through the shell.

Terebratulina is found around the North Atlantic coastline. It lives in clear, fairly deep water away from the littoral zone. It is gregarious, and is attached by a short pedicle to the sea-floor.

Additional common genera

Inarticulate brachiopods

Lingulella (fig.91a) is a small form rather similar to *Lingula* in external form, but tapering more gently towards the posterior end; there is a groove for the pedicle in the pedicle valve.

Cambrian to Middle Ordovician.

Orbiculoidea (fig.91b,c). The shell is circular in outline. The brachial valve is conical and the pedicle valve is almost flat. A furrow in which lies a small pedicle opening, runs from the apex to the posterior margin of the pedicle valve.

Ordovician to Permian.

Crania (fig.91d,e) unlike most inarticulate brachiopods has a calcareous shell. It lacks a pedicle and is cemented by the flat pedicle valve to a hard surface, often occurring on other shells. The brachial valve is conical.

Cretaceous to Recent.

Articulate brachiopods

Orthis (fig.91f–h) has a biconvex shell of almost semicircular outline with a straight hinge line which is not the widest part of the shell. The pedicle valve has a narrow interarea, with an open delthyrium and there is a similar opening in the hinge line of the brachial valve. The surface of the shell is traversed by radial ribs or striae. Inside the pedicle valve the teeth are supported by dental plates. There are crura in the brachial valve.

Ordovician.

Pentamerus (fig.91j–l) has a large smooth shell of oval outline. Both valves are highly convex and the beaks are strongly incurved over the hinge line which is short and curved. There is no interarea. The delthyrium is wide and there may be a deltidium. Inside the pedicle valve dental plates form a spondylium which is supported on the floor of the valve by a double septum. The spondylium is often seen in fossils since these tend to break readily along the internal plates.

Silurian.

Conchidium (fig.91i) is similar to *Pentamerus* but is more globose and has a surface ornament of strong radial ribs.

Upper Ordovician to Silurian.

Leptaena (fig.92a–d). The shell is almost rectangular in outline with a straight hinge line which forms the widest part of the shell. The posterior part of the shell is flat, and the anterior part is bent abruptly almost at right angles to it in such a way that the pedicle valve is convex, and the brachial valve is concave. The surface of the shell is covered with radial ribs and these are crossed by concentric corrugations on the flat part of the shell. Each valve has an interarea. There is a deltidium in

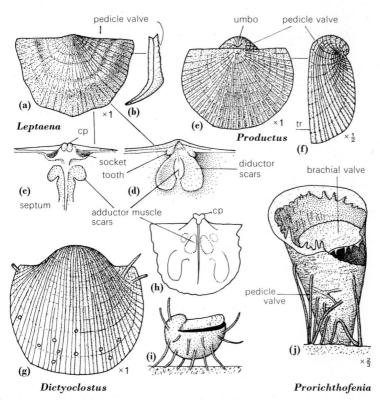

92   A strophomenid and productoids.

a–d, strophomenid; a, pedicle valve; b, lateral view; c, inside of the posterior end of the brachial valve and d, of the pedicle valve (cp, cardinal process). e–j, productoids; e, brachial valve; f, side view of a specimen with the trail (tr) preserved; g, pedicle valve; h, inside of the posterior end of the brachial valve; i, side view of a spinose shell to show its presumed life attitude, supported by its spines off the sea-bed. j, an aberrant productoid cemented to the sea-bed and anchored by spines; the pedicle valve is broken to show the position of the lid-like brachial valve within the pedicle valve.

the pedicle valve, and a similar plate in the brachial valve. The foramen
is very small.

Ordovician to Devonian.

Productus (fig. 92e, f) is semicircular in outline and has a long straight
hinge line. The pedicle valve is strongly convex with a broad rounded

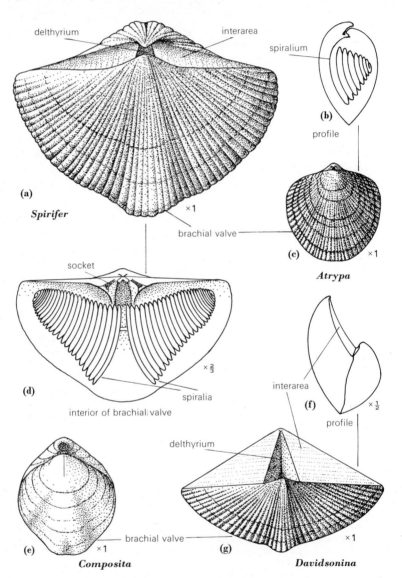

93   Spire-bearing brachiopods.

umbo. The posterior part of the brachial valve is almost flat, but the anterior part is bent almost at right angles to the earlier part so that it lies parallel and very close to the pedicle valve. This anterior part of the shell is called a 'TRAIL'; it is often broken away in the fossil giving a misleading appearance that this is complete as in fig.92e. Close-set ribbing radiates from the umbones, and there are concentric corrugations on the posterior part of the shell. Long hollow spines were dispersed irregularly over the surface of the shell in life, but these are seldom preserved and their bases are marked by small projections from the surface. The spines were apparently a device to anchor the shell on a soft sea-floor (fig.92i). There is no delthyrium. *Productus* and many related genera, e.g. *Dictyoclostus* (fig.92g) are abundant in the Lower Carboniferous where they are of considerable stratigraphic importance.

Carboniferous

Richthofenia is an aberrant member of the *Productus* group, with a cone-shaped pedicle valve which was cemented by the apex to the sea-floor and buttressed by an irregular outgrowth of spinose rooting processes. The brachial valve forms a flat lid. A related genus, *Prorichthofenia* (fig.92j) has spines around the aperture and the brachial valve is recessed within the pedicle valve. The richthofeniid shell, like that of the rudists (p.51) resembles in shape the corallum of a solitary rugose coral. These brachiopods lived in closely clustered communities in a reef-like environment.

Permian.

Spirifer (fig.93a) has a roughly triangular biconvex shell with the straight hinge line forming the widest part. There is a deep median sulcus in the pedicle valve and a corresponding fold in the brachial valve. The umbo of the pedicle valve is prominent, and under it is a large interarea interrupted by the delthyrium. This is partly closed by the deltidium which surrounds the foramen. The interarea of the brachial valve is narrow. The ornament consists of numerous radial ribs. In the interior there are brachidia in the form of spiralia which are attached to crura. The axes of the spiralia lie parallel to the hinge line with the apices directed outwards (fig.93d).

Ordovician to Permian.

Other spire-bearing genera shown in fig.93 include *Atrypa* (Silurian to Devonian), *Davidsonina* (Lower Carboniferous) and *Composita* (Upper Devonian to Permian).

Rhynchonella is a small stout form, almost triangular in outline, and tapering towards the posterior end. The beak of the pedicle valve is sharp and projects above the short curved hinge line. The foramen lies under the beak, and there are small deltidial plates. The surface may be smooth or may have radial ribs or costae; there is a strong fold in the brachial valve, and a corresponding deep sulcus in the pedicle valve. Internally the teeth are supported by dental plates. Narrow crura curve towards the pedicle valve, and there is a small septum on the floor of the brachial valve.

Rhynchonella s.s. Upper Jurassic.

There are many forms ranging in age from the Ordovician onwards which bear a general resemblance to *Rhynchonella*; these may be referred to *Rhynchonella* in a broad sense (*sensu lato*) or grouped as rhynchonellids. They show external differences in some cases, and also internal differences in the pattern of the muscle scars. On the basis of these differences they are divided into separate genera. Some examples are *Sphaerirhynchia* [*Wilsonia*], Silurian (fig.94f), *Pugnax*, Devonian and Carboniferous (fig.94e), *Tetrarhynchia*, Jurassic (fig.94a–c), *Cyclothyris*, Cretaceous, (fig.94d).

Terebratula (fig.94g), has an elongate-oval shell which tapers towards the posterior end. Both valves are convex. The anterior margin is gently folded. The surface is smooth, and ornamented with fine growth lines. The foramen is large and pierces the beak; under it lie the deltidial plates. The brachidium consists of a short loop.

Tertiary.

'*Terebratula*' is used in a broad sense to include a variety of loop-bearing brachiopods. A large number of genera can be distinguished, partly by external details but largely by internal differences especially in the form of the loop. Broadly speaking two groups may be distinguished: (i) forms with a short loop like *Terebratula* s.s., Tertiary, *Dielasma*, Carboniferous, *Epithyris*, Middle Jurassic (fig.94h), *Dictyothyris*, Middle Jurassic (fig.94i), *Terebratulina*, Jurassic to Recent, (fig.90g); (ii) forms with a long loop, like *Ornithella*, Middle Jurassic (fig.94j), and *Stringocephalus*, Devonian.

Geological history of the brachiopods

SUMMARY. Brachiopods appeared in the early Cambrian. Their numbers increased greatly during the Lower Palaeozoic to a maximum in the Silurian and Devonian, but declined gradually during the Upper

Palaeozoic. Only a few stocks survived into the Mesozoic of which two groups are at times very common. They are less abundant in the Tertiary, and are not generally common today (68 genera).

Inarticulate brachiopods are more common in the Cambrian than articulate forms. They are mainly small rounded or oval forms, e.g. *Lingulella* (fig.91a). After the Cambrian inarticulate brachiopods are rarely common apart from the sporadic occurrences of certain long ranged forms with a conservative shell morphology, like for instance *Lingula*.

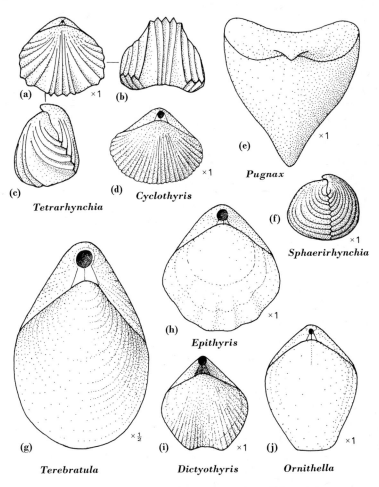

94   Rhynchonellids and terebratulids.

a–f, rhynchonellids; in b, note the strong fold in the anterior margin of the shell; in (e) the shell is slightly tilted to show more of the pedicle valve. g–j, terebratulids.

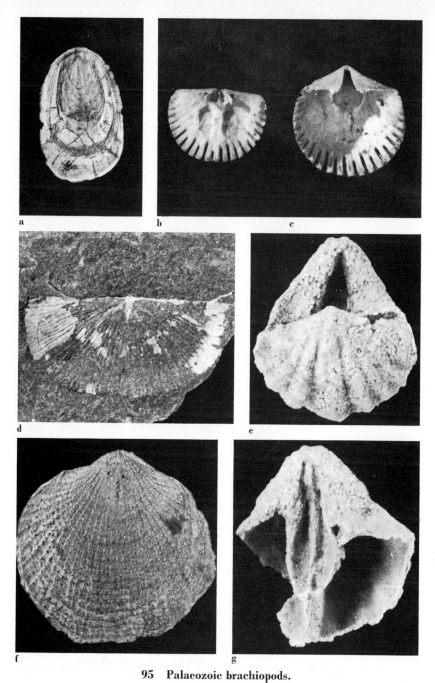

95   Palaeozoic brachiopods.

a, *Lingula squamiformis*, Lower Carboniferous (×2). b, c, an orthid, *Hesperorthis*, Middle Ordovician (×2). d, a strophomenid, *Chonetes*, Silurian (×2). e, g, a spiriferid with spondylium, *Cyrtina carbonaria*, Carboniferous Limestone (×4). f, *Atrypa reticularis*, Silurian (×2).

Articulate brachiopods are found in considerable numbers in the Ordovician and during the Palaeozoic they are often of stratigraphic importance especially in shelly facies, e.g. in the Caradocian. *Orthis* (fig.91h) and related forms (fig.95b,c) are the most characteristic brachiopods of the Ordovician, but most of the main groups are also represented. Brachiopods are most numerous, and also most diverse in form during the Silurian and Devonian. *Leptaena* (fig.92a), *Atrypa* (fig.95f), *Conchidium* (fig.91i), and *Pentamerus* (fig.91l), are some of the commonest Silurian brachiopods. *Spirifer* (fig.93a) and forms related to *Leptaena* (strophomenids) may be cited as common Devonian examples. There was a general decline in numbers and variety of genera in the Carboniferous and Permian. The dominant forms of this time are related to *Productus* (fig.92e) and *Spirifer* (see fig.95e, g). *Productus* has been subdivided into a large number of genera which can be distinguished readily by surface ornament and external form. Productoids are common in the Carboniferous Limestone in England and they are useful stratigraphically especially in limestone facies.

Productoids and spirebearing forms are also the dominant Permian brachiopods. They occur in the Magnesian Limestone in England. One of the best known fossils from this horizon is the highly spinose ' *Productus*' *horridus*. *Prorichthofenia* (fig.92j) with its unusual morphology is an interesting Permian productoid. It occurs in a marine reef-like facies common in eastern areas from Sicily through India to China, but is absent from the marine Permian in England.

	INARTICULATES	ORTHIDS	PENTAMERIDS	STROPHOMENIDS	PRODUCTIDS	SPIRIFERIDS	RHYNCHONELLIDS	TEREBRATULIDS
QUATERNARY								
TERTIARY								
CRETACEOUS								X
JURASSIC							X	
TRIAS								
PERMIAN					X			
CARBONIFEROUS					X			
DEVONIAN						X		
SILURIAN			X	X		X		
ORDOVICIAN	X							
CAMBRIAN								

96 Geological distribution of the main groups of brachiopods.
X marks the period of time during which a group reached its acme.

A number of different stocks of brachiopods survived into the Mesozoic, including spirebearing forms. The most commonly occurring forms however, are rhynchonellids and terebratulids. The rhynchonellids are in general less important after the Jurassic; examples are *Tetrarhynchia*, Jurassic (fig. 94a) and *Cyclothyris*, Cretaceous (fig. 94d). The terebratulids are the dominant Mesozoic brachiopods, and remain so throughout the Tertiary and today. Examples include *Terebratula*, Tertiary (fig. 94g), *Ornithella*, Middle Jurassic (fig. 94j), and *Dictyothyris*, Middle Jurassic (fig. 94i).

TECHNICAL TERMS

ADDUCTOR MUSCLES muscles which on contracting, close the valves. They leave a pair of muscle scars in the pedicle valve (fig. 89e), and two pairs in the brachial valve (fig. 89f).

ANTERIOR MARGIN the margin of the shell along which the valves open.

BEAK pointed tip of a valve at its posterior end.

BRACHIAL VALVE the dorsal valve to which the brachidium or the lophophore is attached; it is usually the smaller valve (fig. 89f).

BRACHIDIUM a calcareous support for the lophophore in some groups of articulate brachiopods. It is usually loop-shaped (fig. 89f) or spiral (fig. 93d); see also crura.

CARDINAL PROCESS a knobbly projection from the hinge line of the brachial valve to which the diductor muscles are attached (fig. 89f).

CRURA (crus) a brachidium in the form of two short calcareous prongs (fig. 91f).

DELTHYRIUM a triangular gap along the hinge line of the pedicle valve, through which the pedicle emerges (fig. 89e).

DELTIDIAL PLATES two plates which partly or completely fill in the delthyrium (fig. 89a).

DELTIDIUM a single plate which fills in the delthyrium except for the pedicle opening.

DENTAL PLATES plates which project from the floor of the pedicle valve to support the teeth.

DIDUCTOR MUSCLES the muscles which by contracting open the valves. There are two pairs which run from the floor of the pedicle valve to the cardinal process in the brachial valve (fig. 89c).

FORAMEN a circular opening in, or near, the umbo of the pedicle valve through which the pedicle emerges (fig. 89a, see pedicle opening).

HINGE LINE line of articulation between the valves along the posterior margin of the shell.

LOPHOPHORE the feeding mechanism consisting of a pair of grooved lobes or arms (brachia) with a fringe of ciliated tentacles (fig. 89d).

MANTLE two extensions of the body wall which enclose the viscera and secrete the valves (fig. 89d).

MUSCLE SCARS the impressions on the inside of the valves which mark the areas of attachment of the adductor, diductor and pedicle muscles (fig. 89e, f).

PEDICLE the stalk by which the brachiopod is attached to the sea-floor (fig. 89d).

PEDICLE OPENING the opening in the pedicle valve, or between the two valves through which the pedicle passes (see foramen).

PEDICLE VALVE the ventral valve to which the pedicle is attached (fig. 89e).

SOCKET one of two pits in the hinge line of the brachial valve into which the teeth of the pedicle valve fit (fig. 89f).

SPONDYLIUM a V-shaped arrangement of plates in the pedicle valve, to which muscles were attached (fig. 91j, k).

SULCUS a downfold of the anterior region of one valve opposite a complementary upfold, FOLD in the other valve.

TOOTH one of two projections along the hinge line of the pedicle valve which fit into sockets in the hinge line of the brachial valve (fig. 89e).

UMBO the rounded portion of either valve around the beak (fig. 92e).

12

Coelenterata

The Coelenterata contain the simplest of the many-celled animals (Metazoa) in which definite tissues are developed. Examples include the jellyfish and sea-anemones which are soft bodied, and the reef-building corals which secrete a calcareous skeleton; there are also a variety of extinct corals.

Coelenterates are exclusively aquatic and the majority are marine. They may be sessile (sea-anemone) or free swimming (jellyfish). The sessile form, or POLYP, may occur as a single individual (SOLITARY) or be united with others to form a colony (COMPOUND). It is sac-shaped, and is attached at its base to the sea-floor with its one opening, the mouth, at its upper free end. The free-swimming form, or medusa, resembles an umbrella with tentacles hanging down round the margin. Some coelenterates like corals exist only as polyps, while in others the life history involves an alternation of polyp and medusoid generations, the one giving rise to the other.

The coelenterate body is of simple structure consisting of an outer layer, ectoderm, and an inner layer, endoderm, which in more advanced members like corals are separated by a jelly-like substance. The body wall encloses a central cavity, the gut or COELENTERON, and this may be divided by radial partitions, the mesenteries, which aid in digestion and absorption of food. The mouth serves for both intake of food, and discharge of waste and larvae. It is surrounded by retractile tentacles which are armed with nematocysts, stinging organs almost confined to coelenterates. There is no blood system, and the nervous system is a diffuse network of cells. Reproduction is sexual or asexual. Typically the body shows radial symmetry, the parts of the body being repeated around the mouth, but in some forms there is also bilateral symmetry.

Coelenterates are commoner in warm shallow seas, but some forms live at depths down to 6000 m, and in temperatures as low as 1 °C.

Simple medusoid fossils have been found in rocks of late Pre-Cambrian age. Apart from these, fossil coelenterates belong to one of three classes: the Scyphozoa (or jellyfish); the Hydrozoa (or hydroids) which includes the stromatoporoids; and the Anthozoa (or corals and sea-anemones). Only the corals, however, are common as fossils, and need be considered in detail.

CLASS ANTHOZOA (CORALS)

The individual anthozoan is a polyp with the coelenteron divided by six, eight or more radial mesenteries and with a gullet between the mouth and the coelenteron. Three orders are important, a surviving one, the Scleractinia (Hexacorallia) or reef-building corals and two extinct groups, the Rugosa and the Tabulata. They are all marine groups.

The various groups of fossils dealt with in earlier chapters have been simple to interpret because there are surviving forms which are more or less nearly related. This is also true of the corals, and in particular the scleractinian corals, which are widespread as fossils in the Mesozoic and provide the basis for our knowledge of fossil corals in general. The rugose corals, however, which are geologically the more important and which show a greater variety of skeletal structure, are described first.

RUGOSA

The rugose corals are solitary or compound Palaeozoic corals with a calcareous skeleton, the CORALLUM, divided by vertical radial partitions, the SEPTA. These include six main PRIMARY septa together with a number of later formed septa which during growth were inserted at four points in the corallum.

Rugose corals are common fossils in Britain, especially in the Lower Carboniferous rocks where they are of considerable stratigraphic value. Their structure is relatively simple and can be examined in detail either by grinding down the specimen with an abrasive or by cutting through it at various points with a rock-saw.

Morphology

The CORALLUM is bounded on the outside by a wall which may be transversely wrinkled, a feature which gives rise to the term 'rugose'. Its upper surface is a cup-shaped hollow of varying depth, the CALICE (fig. 97c), and this by analogy with modern corals is the surface on which the rugose polyp 'sat'. The skeleton consists for the most part of vertical elements, the SEPTA; smaller convex plates, the DISSEPIMENTS, which lie between the septa; and transverse elements, the TABULAE, which are flattish plates.

In a solitary rugose coral the corallum is basically conical in shape (the tip being the first formed part), and may be straight or curved (horn-shaped, fig. 97c); the later-formed part may be cylindrical (fig. 97g). In compound forms the corallum is built up of a number of individuals, each of which is referred to as a corallite. These may remain free, FASCICULATE (fig. 97e, f) or may be in contact, MASSIVE (fig. 97d). A fasciculate corallum is further described as DENDROID if the corallites branch irregularly (fig. 97e), and as PHACELOID if the corallites are more or less parallel (fig. 97f). A massive corallum is described as

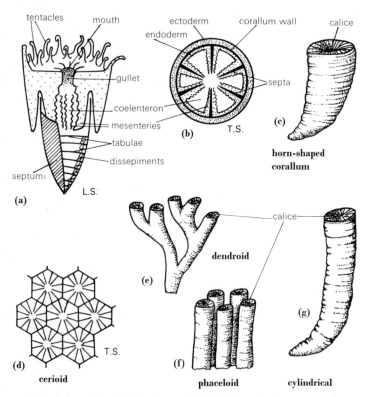

97   Morphology of the rugose corals.

a, longitudinal section (L.S.) of a hypothetical corallum and polyp. b, transverse section (T.S.) to show the probable relationship between the skeleton and soft tissues. c, g, solitary corals; d–f, compound corals; d, massive; e, f, fasciculate.

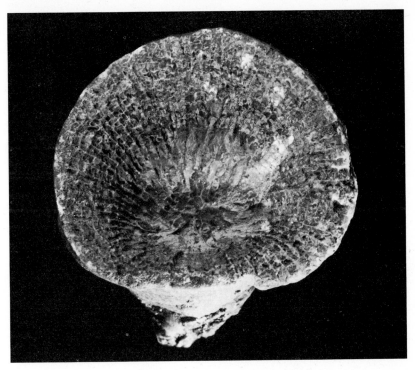

98   Calice of a solitary rugose coral.

CERIOID if the corallites are polygonal in shape (transverse section) and are united by their walls (fig. 97d), or as ASTRAEOID if the corallite walls are lacking (fig. 103d).

The CALICE (figs. 97c, 98) may be a shallow or relatively deep depression; its centre is the AXIAL REGION. In a small number of forms it could be closed by a lid (OPERCULUM, fig. 103c). The septa project from the floor of the calice, and presumably they extended between pairs of mesenteries (p. 160) in the living polyp.

The growth of the septa can be traced by cutting the corallite transversely at intervals along its length. The first formed are six PRIMARY SEPTA; these are named as the CARDINAL, COUNTER, two ALAR and two COUNTERLATERAL septa and lie as shown in fig. 99a, b. Subsequent septa are added at four points, i.e. on each side of the cardinal septum, and on the counter side of the alar septa. In some Rugosa there are relatively conspicuous spaces, FOSSULAE, at the points where new septa are inserted. The most obvious of these is the cardinal FOSSULA which is readily identified by the (usually) short cardinal septum which lies in it (fig. 99c). Forms in which the primary septa can be distinguished show bilateral symmetry, but in many corals, especially colonial forms, the septa are very numerous in the adult stages and develop a dominantly radial symmetry with alternating longer, MAJOR, and shorter, MINOR septa (fig. 99d).

TABULAE (fig. 99h) are flat or gently domed plates which represent the floors of successive calices. They were secreted at the base of the polyp to seal off the earlier untenanted parts of the corallum. Tabulae are most clearly seen in a longitudinal section, especially in corals in which the septa are short. In corals with fossulae, the tabulae are folded

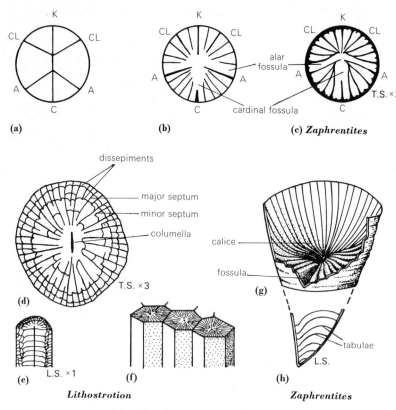

99   Morphology of rugose corals.

a, b, transverse sections of a solitary rugose coral, not to scale; a, juvenile stage with the six primary septa (C, cardinal septum; A, alar septum; K, counter septum; CL, counter-lateral septum); b, adult stage with later-formed septa added. c, g, h, *Zaphrentites*; c, transverse section; g, upper part of a corallum, partly broken, to show the deep calice; h, lower part of a corallum sectioned to show the tabulae. d–f, *Lithostrotion*; d, e, a single corallite of a phaceloid species; f, a cerioid species.

100   Growth of new individuals in a compound coral. Four corallites have developed within the axial region of the topmost calice. *Acervularia,* **Wenlock Limestone, Silurian (×4).**

into trough-shaped hollows which slope down to the walls in the vicinity of the fossulae. Forms with minor septa may have a peripheral zone in which there are small arched plates, DISSEPIMENTS, instead of tabulae (fig. 99e).

While the axial region may remain a clear space, in many forms there is some sort of axial structure. This may take the form of a vertical rod, the COLUMELLA (fig. 99d), or a more complex structure (fig. 103e) consisting of plates, some radial, and others concentric about the axis of the coral.

INCREASE OF CORALLITES. The number of individuals in a compound coral is increased by a vegetative process, either by splitting (fission) of a parent polyp or by the growth of buds (budding) from the parent polyp. In the latter case the new polyps may grow out from the side of the parent, or from within the calice, as in fig. 100, where four corallites have developed in the axial region of the calice, a process known as axial increase.

101   *Lithostrotion, Carboniferous Limestone*, a thin section of a cerioid species. Above, a transverse section ($\times 3\frac{1}{2}$). Below, a longitudinal section ($\times 5\frac{1}{2}$).

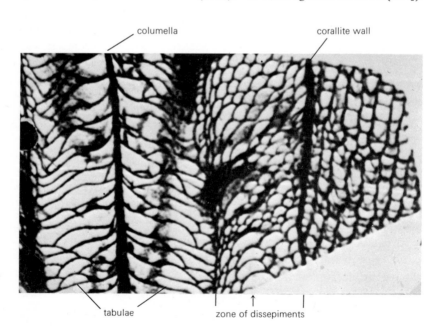

102   *Lithostrotion*: a cerioid corallum (×2).

Skeletal form in two common rugose corals

Zaphrentites (fig. 99c, g, h) is a solitary form. It has a horn-shaped corallum with a deep calice. There is marked bilateral symmetry; the cardinal and countercardinal septa lie in the plane of symmetry and the other septa slope towards this plane. The inner end of the major septa often fuse in the axial region. There are usually no minor septa. The cardinal fossula is an elongated space bounded laterally by major septa, and the short cardinal septum projects into it. There are no dissepiments. The tabulae are rather irregular, and inversely conical in shape.

Lower Carboniferous. Related genera range from the Carboniferous to the Permian.

Lithostrotion (figs. 99d–f and 101, 102) is a compound coral. In some species the corallum is phaceloid (fig. 99e), in others it is massive and cerioid (figs. 101, 102). There are major and minor septa, radially arranged; the major septa are long and may unite with a rod-like columella. The columella may be absent in some individuals. There is usually a narrow zone of dissepiments. The tabulae are arched; they are flat in corallites lacking a columella.

Lower Carboniferous.

Mode of life of rugose corals

The clues to the ecology of rugose corals lie in the type of rock in which the fossils are found, the associated fossils and, of course, in analogy

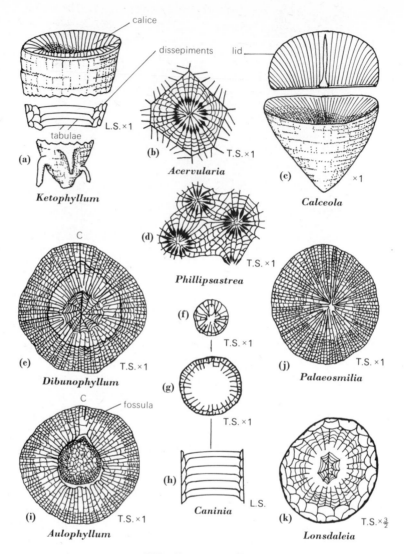

103 Rugose corals.
a, c and e–j, solitary corals. b, d, k, compound corals; in (k) one individual only is shown.

with modern corals. Some Rugosa occur mainly in calcareous rocks. *Lithostrotion* and *Lonsdaleia* (p. 170) for instance often form extensive sheets which can be traced for many miles in the Lower Carboniferous of the North of England. The enclosing rock is a light-coloured limestone in which brachiopods are next in abundance to the corals, and it formed, perhaps, in shallow and relatively warm water. Some of the small horn-shaped.solitary corals like *Zaphrentites* occur in limestones and also in dark grey shales which were formed in deeper, quieter conditions; they represent an environment from which the colonial Rugosa are conspicuously absent.

Additional common genera

Solitary Rugosa

Ketophyllum [*Omphyma*] (fig. 103a) has a conical corallum with a deep calice. The tabulae are flat in the axial region and are replaced by dissepiments towards the periphery. The septa are discontinuous. There are rootlike outgrowths from the base.

Silurian.

Calceola (fig. 103c). The corallite is semicircular in transverse section. The calice is very deep, and it has a lid (operculum). The septa form fine ridges in the calice.

Devonian.

Caninia (fig. 103f–h). The initial part of the corallite is horn-shaped, and the major septa extend into the centre; the adult part is cylindrical and the major septa are shorter and do not reach the axial region. The cardinal fossula is small and open. The minor septa are very short. A small number of irregularly arranged dissepiments occur in the periphery of the corallite. The tabulae are flat in the axial region but are turned down towards the margins.

Carboniferous to Permian. In England *Caninia* occurs in the Lower Carboniferous.

Dibunophyllum (fig. 103e). The corallite is cylindrical, except at the base. There is a small open cardinal fossula readily distinguished by the short cardinal septum. The major septa are numerous and extend almost to the axial structure. Minor septa are irregularly developed and there is a wide zone of closely spaced dissepiments. The axial structure resembles a spider's web with concentric and radially arranged plates; the most prominent radial plate lies in the plane of symmetry. Tabulae are confined to the area between the zone of dissepiments and the axial structure.

Lower Carboniferous. *Dibunophyllum* is the characteristic coral of zone D in England.

Aulophyllum (fig. 103i) resembles *Dibunophyllum* from which it may be distinguished by the form of its axial structure. This consists of a haphazard arrangement of radial and concentric plates surrounded by an outer wall; this wall, seen in transverse section, has a sharp projection towards the cardinal fossula.

Lower Carboniferous.

Palaeosmilia (fig. 103j) is normally solitary, but one species forms small colonies. The solitary corallite is long, and cylindrical. The septa are very numerous; the major septa extend to the axial region where their ends may fuse, the minor septa extend about half-way to the centre. The cardinal fossula is narrow, and not easy to distinguish. There is a wide zone of abundant dissepiments. The domed tabulae are incomplete.

Lower Carboniferous.

Compound Rugosa

Lonsdaleia (figs. 103k, 104). The corallum is fasciculate in some species, and massive and cerioid in other species. The septa major and minor, are separated from the corallite wall by a zone of dissepiments, and from the centre by an axial structure of spider's web pattern. The tabulae are gently curved.

Lower Carboniferous.

Phillipsastrea (fig. 103d) has a massive astraeoid corallum. The inner ends of the septa are thickened.

Devonian.

Acervularia (fig. 103b) has a massive cerioid corallum. The inner ends of the septa are thickened and form a wall round the axial region.

Silurian.

Geological history of the rugose corals

SUMMARY. The Rugosa appeared in the Middle Ordovician, but first became common in the Silurian; their numbers increased gradually to a peak in the Lower Carboniferous and they disappeared in the Permian.

In the Silurian, the Rugosa were minor associates of the tabulate corals (p. 176) and the stromatoporoids (p. 181) in clear, shallow-water limestones like the Wenlock Limestone; e.g. *Ketophyllum* (fig. 103a) and *Acervularia* (fig. 103b). Many new forms appeared in the Devonian; *Calceola* (fig. 103c) and *Phillipsastrea* (fig. 103d) were common examples.

The Rugosa were most numerous in the Lower Carboniferous. New families appeared, replacing some of the earlier ones. Some forms had

104 *Lonsdaleia*, upper surface, Carboniferous limestone ($\times$4).

complex axial structures which were lacking in Pre-Carboniferous Rugosa. The Rugosa are of considerable stratigraphic importance in the Lower Carboniferous and they provide three out of five of the original choice of zonal indices. These are *Dibunophyllum*, D (fig. 103e), *Caninia*, C (fig. 103g) and *Zaphrentis*, now called *Zaphrentites*, Z, all solitary corals. Compound Rugosa, e.g. *Lithostrotion* (fig. 101) and *Lonsdaleia* (fig. 104) are found extensively at some levels in the upper part of the Lower Carboniferous (Viséan).

The Rugosa are not found in England after the Lower Carboniferous. They occur elsewhere in marine rocks of Upper Carboniferous and Permian age but they became extinct at the end of the Permian.

It is generally thought that a group of rugose corals provided the stock from which the Scleractinia developed in the Trias. The two orders show a number of structural similarities, but a major point of difference is the way in which new septa are inserted.

SCLERACTINIA

The Scleractinia are solitary or compound corals with a calcareous skeleton in which septa form cycles of six and multiples of six.

Morphology

The coral skeleton, the CORALLUM, is secreted by the outer wall of the body and so is an exoskeleton. It is formed by the fusion of minute fibres of aragonite. Each polyp sits in a cup-shaped structure, the CALICE, from which the SEPTA project between the pairs of mesenteries which partition the coelenteron.

SOFT BODY. Coral polyps may show some diversity in shape, especially in compound forms where the adjacent polyps may be connected by common soft tissue. A solitary polyp is basically like a sea-anemone,

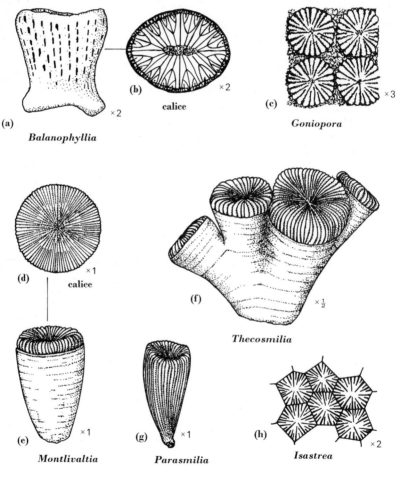

(a) Balanophyllia ×2

(b) ×2 calice

(c) ×3 Goniopora

(d) ×1 calice

(e) ×1 Montlivaltia

(f) ×½ Thecosmilia

(g) ×1 Parasmilia

(h) ×2 Isastrea

105 Scleractinian corals.
Solitary corals, a, b, d, e, g. Compound corals, c, f, h.

sac-shaped with an oval mouth surrounded by one or more rings of tentacles. The tentacles are armed with nematocysts which are partly for defence, and partly are used to paralyse small animals. These are then grasped as prey by the tentacles and thrust into the mouth. Food passes from the mouth through the gullet to be digested in the coelenteron and waste is ejected through the mouth. The mesenteries radiate in pairs from the wall of the coelenteron. There are six primary mesenteries and a varying number formed later.

SKELETON. The corallum of a solitary coral is commonly conical or cylindrical in shape. The corallum of compound corals shows a wide range in shape and size, especially in the modern reef corals. The terms used to describe the form of the corallum have been defined in the section on rugose corals (pp. 161–165).

The first formed part of the corallum is a basal plate by which it is attached to the sea-floor. On this, the polyp secretes the series of vertical and transverse plates from which the corallum is built up. The polyp is in contact with the upper surface only, of its corallum, the calice, and this is in effect a mould of the base of the polyp. From time to time as the polyp outgrows its calice, it moves upwards and secretes a new calice.

The corallum is typically simpler in structure in the scleractinian than in the rugose coral. Septa are the dominant skeletal elements. There are six primary septa. These are the longest and may mark the corallite into sextants. Later septa are inserted in CYCLES, each comprising a multiple of six septa. There are usually no more than four or five cycles. A rod or plate-shaped COLUMELLA may be present, but more elaborate axial structures are uncommon. DISSEPIMENTS occur between the septa. They are formed during growth by the polyp to cut off the lower part of the corallum from which it has withdrawn.

Skeletal form and mode of life of two scleractinian corals

Balanophyllia (fig. 105a, b) is a solitary coral. The corallite is cylindrical and is attached to rocks by a broadened base. The septa are crowded, and their inner edges are partly fused; the longest septa join a columella of rather spongy texture. *Balanophyllia* occurs in the Red Crag in Norfolk.

Tertiary to Recent.

Balanophyllia is found in rock pools on the south-west coast of England, and occurs down to about 1000 m. It is a 'cup' coral resembling a brilliant scarlet sea-anemone with yellow tentacles, until it contracts within its calice and reveals the skeleton. Similar cup or 'deep sea' corals may be found in deep, relatively cold, water. Some are colonial but form only small colonies.

Goniopora (fig. 105c) is a massive colonial coral; the individual corallites are polygonal in transverse section. The septa are arranged in three cycles, and have serrated edges; the larger septa unite in a columella. *Goniopora* is found in the Eocene (Bracklesham Beds) in England (fig. 108).

Middle Cretaceous to Recent.

Goniopora is one of the reef-forming corals which today build reefs, often of very great extent, in the tropical and subtropical zones between latitudes $35°N$ and $32°S$. Reef corals will not tolerate a wide range in salinity, temperature or depth. They require a salinity of about 36 parts per thousand, a temperature between about $18 °C$ and $29 °C$ and a depth of not more than 90 m. They grow most luxuriantly in sediment-free, and well-oxygenated water between $25 °C$ and $29 °C$. These stringent requirements are due to the presence in the polyp walls of unicellular algae (Zooxanthellae). The algae have a symbiotic relationship with the polyp. (Symbiosis is an association of mutual benefit between unlike organisms.) The algae require light and warmth as well as carbon dioxide and nutrient salts for photosynthesis. Thus the host corals are restricted to the photic zone (p. 16). The hosts benefit in turn by a supply of oxygen, the removal of CO_2 and other waste products, and perhaps by food. The algae leave no trace of their existence in the coral skeleton.

Additional common genera

Montlivaltia (fig. 105d, e) is solitary; the corallum is broadly conical and many septa project into the shallow calice. The columella is absent or poorly developed; there are dissepiments. *Montlivaltia* belongs to the reef-forming group.

Mesozoic.

Thecosmilia (fig. 105f) is related to *Montlivaltia*; it forms small colonies.
Mesozoic.

Isastrea (fig. 105h) is a colonial coral with a massive cerioid corallum. The corallites are hexagonal or pentagonal. There are about four cycles of septa; dissepiments are present. *Isastrea* is a reef-forming coral.
Middle Jurassic to Cretaceous.

106 A coral reef.
(Underwater photograph of the Key Largo Coral Reef Preserve, Florida Keys.)

Parasmilia (fig. 105g) is a solitary cup coral which is fixed by a broadened base. The corallum is somewhat elongate and may be cylindrical in the mature stage. The surface is ridged by many (48) septa which are arranged in four cycles. A columella stands up prominently in the calice.

Cretaceous to Recent. It is common in the Chalk.

Geological history of the scleractinian corals

SUMMARY. The Scleractinia appeared in the Middle Trias and assumed importance in the Jurassic; they remain abundant today.

The Scleractinia are represented in Britain in the earlier Jurassic by only a few genera, usually solitary forms, e.g. *Montlivaltia* (fig. 105e). Reefs with forms like *Isastrea* (fig. 107) occur later, for instance in the Corallian at Upware. This particular occurrence lies much further north than reefs occur at the present day. Corals are not an important part of the Cretaceous fauna in England and they are mainly solitary cup corals

**107   A Jurassic reef-building coral. *Isastrea*, Corallian, Upware, Cambs.
(× 4). This is an external mould of the upper surface, and the calices appear
in inverted relief.**

like *Parasmilia*, Chalk. Apart from small specimens of *Goniopora*
(fig. 108), a reef-building form occurring in the Eocene (Bracklesham
beds only), only cup corals are found in later deposits in Britain, e.g.
Balanophyllia (fig. 105a), in the Red Crag.

At the present day corals play an important part in reef-formation
within the tropical and subtropical regions mainly around oceanic islands
and along east coasts of larger land masses, e.g. the barrier reefs of
Australia.

TABULATA

The Tabulata are extinct compound corals in which the corallum is
built up of slender corallites partitioned transversely by many tabulae.

Morphology

The CORALLUM is calcareous. It is usually small, perhaps a few centimetres in diameter, but it can be as much as 2 m across. The corallum

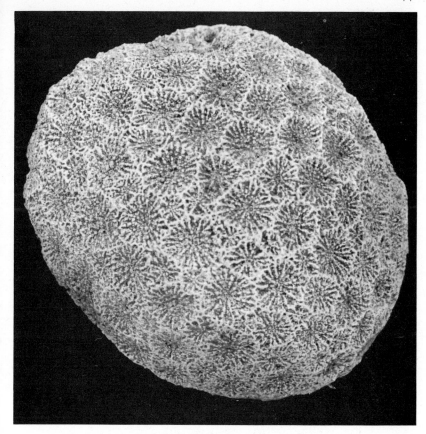

108 A Tertiary reef-building coral. *Goniopora*, Bracklesham Beds, Eocene (×2·6).

may be little more than an incrustation, but more usually it is fasciculate (with individual corallites branching freely) or massive and of irregular bun shape.

The corallites in fasciculate forms may branch irregularly, and in a number of .cases they are united at intervals by connecting tubes (fig. 109h). In a few forms, the corallites unite laterally to form a series of meandering walls one corallite in width (fig. 109c). Some massive forms have communicating holes, MURAL PORES, in the corallite walls (fig. 109b); in others the corallites are separated by a zone of common calcareous tissue, COENENCHYME, in which additional tubes which are slender editions of the corallites may occur (fig. 109d).

The CALICE is rarely more than a few millimetres across and may be round, oval or polygonal in shape. SEPTA, if present, number about 12. They are alike and in transverse section radiate like short spines from the corallite wall. TABULAE are the most characteristic feature of these corals. They are numerous, and typically horizontal (fig. 109b), but they may be domed or funnel-shaped. An axial structure is only exceptionally present.

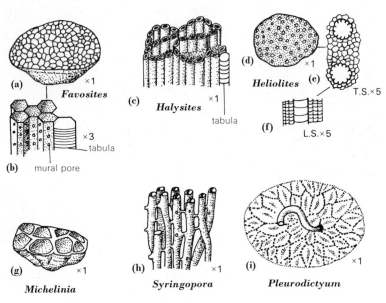

109   Tabulate corals.

Skeletal form in some common tabulate corals

Favosites (fig. 109a, b) has a massive, irregularly hemispherical corallum which typically is only a few centimetres across. It is cerioid, with long slender polygonal corallites which communicate through mural pores. The calice measures about 2 mm in diameter. The septa form ridges inside the walls, or are absent. Tabulae are numerous and evenly spaced. Upper Ordovician to Devonian. Important in the Silurian.

Heliolites (fig. 109d–f) is a massive form in which cylindrical corallites are separated from each other by coenenchyme composed of narrower polygonal tubes. There are 12 short spinose septa. The tabulae are regular and more or less horizontal. Silurian to Devonian.

110   A tabulate coral. *Syringopora*, **Lower Carboniferous** (×1·4). **The corallum is silicified.**

Halysites (fig. 109c) has a phaceloid corallum in which slightly compressed corallites are united on two (or three) sides to form a branching series. Its appearance in transverse section has earned *Halysites* the popular name of 'chain' coral. The tabulae are horizontal.
Ordovician to Silurian.

Syringopora (figs. 109h and 110) forms dendroid or phaceloid colonies. The corallites are united at intervals by transverse tubes. Twelve short septa may be present. The tabulae are funnel-shaped.
Silurian to Carboniferous. Important in the Carboniferous.

Additional common genera

Alveolites is a massive or branching form with the corallites inclined outwards so that they appear oval in transverse section.
Silurian to Devonian.

Pleurodictyum (fig. 109i) has a hemispherical corallum with relatively large corallites (about 5 mm diameter). In one species a central coiled

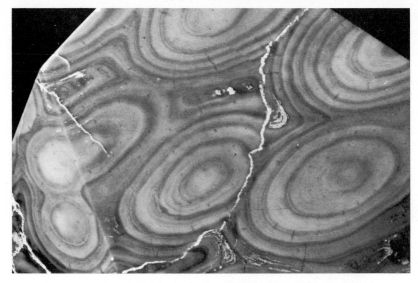

111 A stromatoporoid. *Actinostroma,* **Devonian** (×0·8).

structure is thought to be a worm tube; the association was probably commensal, i.e. neither member greatly influenced the other.
Lower Devonian.

Michelinia (fig. 109g) has a small corallum with relatively large corallites and many convex tabulae, and root-like processes at the base.
Upper Devonian to Permian.

Vaughania is of discoid shape, and has no tabulae. It is the zonal index of the Lower Carboniferous zone K, once known as *Cleistopora.*
Lower Carboniferous.

Geological history of the tabulate corals

SUMMARY. The Tabulata appeared in the Middle Ordovician and were abundant in the Silurian and Devonian. Their numbers decreased in the later Palaeozoic and they disappeared at the end of this era.

The tabulates occur most commonly in rocks of high calcareous content, and are seldom found in muddy sediments. They played a significant role in reef formation especially in Silurian and Devonian times. They outnumbered the rugose corals in the Silurian, *Favosites* (fig. 109a) being the dominant genus, while *Halysites* (fig. 109c) and *Heliolites* (fig. 109d) were also common. *Heliolites* and *Favosites* were still common in the Devonian.

Some of the tabulates are very long ranged, but a number have a restricted range and are of value stratigraphically. Thus *Pleurodictyum* (fig. 109i) is restricted to the Lower Devonian, and, in the Lower Carboniferous, forms like *Syringopora* (fig. 110), *Michelinia* (fig. 109g) and *Vaughania*, the zonal index of K, are useful horizon markers.

STROMATOPOROIDS

The stromatoporoids are extinct colonial organisms generally assigned to the Hydrozoa. The calcareous skeleton (fig. 111) forms an irregular mound or sheet ranging in size up to about 2 m, and shows a closely laminar structure traversed by fine vertical pillars. The group was an important constituent of Silurian and Devonian reefs associated with tabulate corals, compound rugose corals and calcareous algae.

Cambrian to Cretaceous.

TECHNICAL TERMS

ASTRAEOID describes a massive corallum in which there are no walls separating the individual corallites (fig. 103d).

AXIAL REGION the central part of the corallite through which the axis of radial symmetry passes.

AXIAL STRUCTURE any vertical structure in the axial region of a corallite, e.g. columella (fig. 99d). More complicated structures are shown in fig. 103e.

CALICE the cup-shaped depression in the upper surface of a corallite, in which the polyp sat (fig. 98).

CERIOID a massive corallum in which the walls of adjacent corallites are united (fig. 101).

COENENCHYME a zone of common skeletal tissue which may separate individual corallites in a corallum.

COLUMELLA an axial structure in the form of a vertical rod or plate (fig. 99d).

CORALLITE the exoskeleton secreted by an individual coral which may be solitary or one of a colony.

CORALLITE WALL the wall surrounding a corallite.

CORALLUM the entire skeleton of a solitary or of a colonial coral.

DENDROID describes a fasciculate corallum in which the corallites branch irregularly in a tree-like pattern (fig. 97e).

DISSEPIMENTS small convex plates lying between the septa in the periphery of a corallite (fig. 97a).

FASCICULATE describes the corallum of a colonial coral in which the corallites are cylindrical and not in contact.

FOSSULA a larger than usual space between the septa in rugose corals; it is usually associated with one of the primary septa, in particular the cardinal septum (fig. 99c).

MASSIVE the corallum of a colonial coral in which the corallites are in close contact.

MESENTERY one of several radial infoldings of the body wall which forms a vertical partition in the coelenteron.

OPERCULUM a lid covering the calice in some rugose corals (fig. 103c).

PHACELOID a fasciculate corallum in which the corallites are more or less parallel in growth (fig. 97f).

SEPTA (um) radial plates formed on the floor of the calice between the mesenteries and extending vertically through the corallite. The longer septa are called MAJOR septa, and shorter ones, usually alternating with the major, are MINOR septa (fig. 99d).

TABULA a flat or convex transverse plate, which extends (wholly or in part) across the corallite (figs. 97a, 103a).

13

Arthropoda

The Arthropoda are segmented animals with a chitinous external skeleton and paired jointed limbs. In numbers of species, the phylum is the largest of the animal kingdom, and it is also one of the most varied, including such examples as insects, spiders, crabs, centipedes, and most important geologically, the extinct TRILOBITES.

Arthropods are bilaterally symmetrical. The body consists of a number of articulating segments, each bearing a pair of limbs which may have widely different functions. The exoskeleton is flexible over the joints between the segments to allow movement, but over the segments themselves it is rigid. Consequently growth cannot occur except during periodic moulting (ecdysis) when the exoskeleton comes away and a new one is formed. The nervous system is highly developed, with a brain from which ventral nerve cords arise and give off branches to each segment. Blood is circulated in the body cavity by a heart. Eyes, either simple or compound, occur in most forms. Respiration usually takes place by means of gills but may occur through the surface of the body or through tubes (tracheae) ramifying the body. The sexes are separate in most forms.

The arthropods are the dominant living invertebrates. The body organisation is adaptable and members are found in every ecological niche. Some live in water (sea and fresh) others on land; and the largest group, the insects, are capable of sustained flight. There is a general tolerance of extremes in temperature; springtails (wingless jumping insects) are found in freezing arctic conditions, and a variety of forms can withstand heat or the desiccation of deserts. Their remains are found from earliest Cambrian times onwards. Apart from the trilobites, which were abundant in the Lower Palaeozoic, the main classes of arthropods are represented only infrequently in the fossil record and most fossils are of aquatic forms.

Fossil arthropods belong to one of the following classes:

Arachnida: spiders and scorpions
Merostomata: king-crabs and eurypterids
Myriapoda: centipedes and millipedes
Insecta: insects
Crustacea: ostracods, shrimps, and crabs
Trilobita: trilobites

Only the trilobites will be dealt with here in any detail.

CLASS TRILOBITA

The extinct trilobites are assigned to the arthropods by virtue of their segmented bodies with chitinous exoskeletons and paired jointed limbs. They are distinguished by the body being divided longitudinally into three parts, an axial and two lateral regions; by the grouping transversely of the segments into three regions, the CEPHALON, THORAX and PYGIDIUM; and by the form of their two-branched limbs.

The trilobites did not survive the Palaeozoic, and understanding of their body parts and possible behaviour is based largely on the study of modern arthropods like the crustaceans. Trilobites appear to have been entirely marine, living for the most part in shallow waters. They were, on average, small creatures, measuring about 5–8 cm in length and with an overall range in size from about 5 mm to 70 cm.

Morphology

The exoskeleton enclosed and supported the soft body, providing a firm surface of attachment for the muscles and preventing desiccation. It consisted of chitin (p. 8) hardened by impregnation with calcium salts. There is ample evidence that trilobites moulted periodically, so that each one in its lifetime may have discarded many exoskeletons each of which was a potential fossil.

The part usually preserved is the exoskeleton which covered the dorsal and a part of the ventral side of the body. Most of the ventral region appears to have been covered by a soft membrane, and the limbs, which are only exceptionally preserved, by soft chitin.

The dorsal exoskeleton is typically a flattened or gently convex shield with the edges turned under on the ventral side to form a rim, the DOUBLURE (fig. 112e). It is grooved longitudinally by two AXIAL FURROWS which separate an arched central region, the AXIS, from two side regions, the PLEURAE (fig. 112a). The anterior and posterior parts consist of several fused segments forming the head-shield or CEPHALON

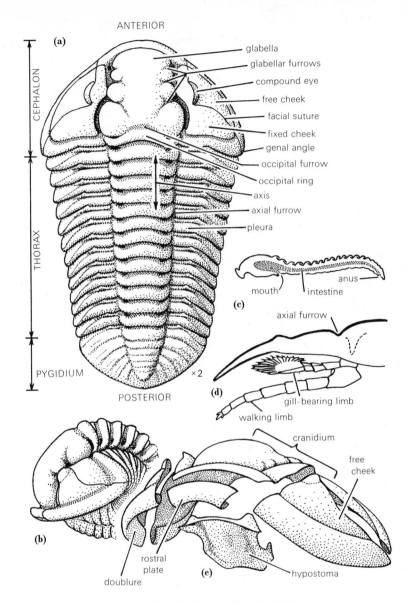

112   Morphology of the trilobites.
a–c, e, *Calymene*; a, dorsal view of the exoskeleton; b, lateral view of an enrolled specimen; c, longitudinal section showing the supposed form of the digestive tract; e, oblique frontal view of the cephalon 'exploded' along the suture lines to show the relationship of the various regions. d, a transverse section of part of a thoracic segment showing a reconstruction of the limbs in *Ceraurus*; anterior view.

113   A spinose trilobite. *Miraspis*, Silurian, Czechoslovakia (×4).

and the tail-shield or PYGIDIUM respectively. Between them is the
thorax containing unfused segments which articulate with one another
so that in many cases the trilobite could roll up, or ENROLL, like a
wood-louse, with the head- and tail-shields pressed together to protect
the soft ventral side of the body (fig. 112b).

The cephalon is typically semicircular in outline. The axial region is
called the GLABELLA and is separated by well defined axial furrows
from the side regions, the CHEEKS (fig. 112a). The glabella varies widely
in size and shape. It is generally strongly convex and in many trilobites
shows its originally segmented character in the form of partial or complete transverse divisions, the GLABELLAR FURROWS. The last of these
is the OCCIPITAL FURROW which defines the posterior segment, the
OCCIPITAL RING (fig. 112a).

The cheeks may be continuous in front of the glabella, or may be
separated by it. In the majority of trilobites each cheek is crossed by a

suture, the FACIAL SUTURE (fig. 112a). It is probable that the cephalon split along this line to facilitate moulting. The facial suture defines two regions, one, the FIXED CHEEK remained attached to the glabella, forming with it the CRANIDIUM, while the other is the FREE CHEEK which separated from the cranidium on moulting (fig. 112e). In trilobites which possess EYES, these lie on the margin of the free cheek against the facial suture, and must accordingly have been freed at an early stage in moulting (fig. 112a, e).

The facial suture follows one of several different courses. Starting from the anterior margin it passes on the inner side of the eye and thence cuts, either the posterior margin, OPISTHOPARIAN condition (fig. 116a) or the lateral margin, PROPARIAN condition (fig. 116g) or in a very few forms the angle (GENAL ANGLE) between the posterior and lateral margins of the cephalon, GONATOPARIAN condition (fig. 112a). The facial suture is not conspicuous in a number of trilobites because it runs along the margin of the cephalon (fig. 116b).

Most trilobites have EYES, but a number of unrelated forms lack these organs and are described as BLIND (fig. 116b). The eyes are usually COMPOUND with a large number of separate lenses which are covered individually or collectively, with a clear cornea (fig. 116g). They are borne on the inner margin of the free cheeks abutting against a raised portion of the fixed cheek, the PALPEBRAL LOBE. In early trilobites, the eyes are of elongate crescentic shape (fig. 116a) and may be connected with the front end of the glabella by a band, the EYE RIDGE (fig. 119b). In later forms they are a more compact kidney shape (fig. 116g).

The genal angles (fig. 112a) are commonly drawn out into GENAL SPINES (fig. 116a) and some genera carry additional spines on the margins of the cephalon or on the glabella (fig. 113).

On the ventral side of the cephalon, there may be two or three small plates. One of these, the HYPOSTOMA (fig. 112e) lies in front of the mouth region, and a second between the hypostoma and the anterior margin. The hypostoma was, in some forms, fused to the cephalon, but in others it separated during moulting.

The number of free segments in the THORAX varies greatly, ranging from 2 to over 40, but may be constant in related genera. The segments are generally similar, except in size, becoming narrower towards the pygidium, so that the outline of the thorax tapers posteriorly. Each segment is divided by the axial furrows into a strongly arched AXIAL RING flanked on each side by flatter PLEURAE (fig. 112a, d). The

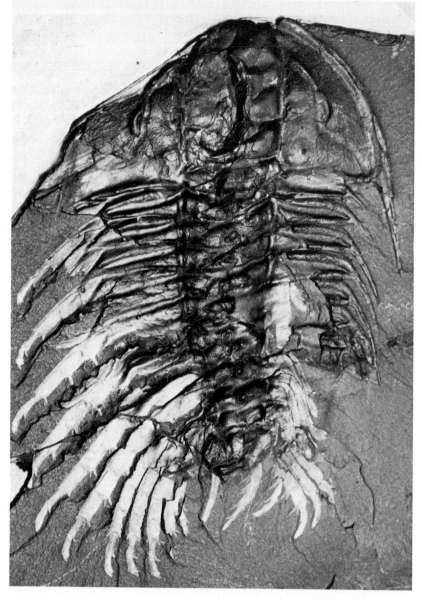

114   A trilobite with some of the appendages (preserved as a thin silvery
film) extending from under the dorsal exoskeleton. *Olenoides serratus*,
Burgess Shales, Middle Cambrian, British Columbia (×1·75).
(Fossil in the United States National Museum.)

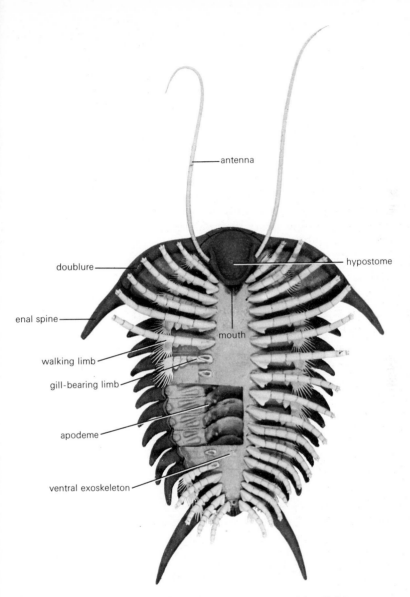

115 Model of the ventral side of a trilobite.

(Based on *Ceraurus* by Mr. F. Munro and Dr. J. K. Ingham.)

pleurae may have rounded or pointed ends. In specimens with the inner surface exposed clear of rock matrix, paired processes (apodemes) to which the limb muscles were possibly attached, may be seen near the axial furrows. (The latter appear as ridges on the inner surface.)

The pygidium is a semicircular or triangular shield which covered a number of fused segments (fig. 112a). These are usually indicated by transverse furrows on both the axial and pleural regions, which give some indication of the number of segments involved, this varying from about 2 to about 30. In some genera, however, the furrows may be partly or wholly effaced, in which case the pygidium may be smooth. The size of the pygidium relative to the cephalon is a useful feature for purposes of classification and the terms MICROPYGOUS, ISOPYGOUS and MACROPYGOUS are used to indicate that the pygidium is smaller, about the same size as, or larger than the cephalon.

Trilobite LIMBS (figs. 112d, 115) have been found in occasional enrolled specimens and in rare fossils preserved in extremely fine-grained sediments (fig. 114). A well-known example is *Triarthrus* from the Middle Ordovician black shale in New York State, U.S.A. Examples of many different ages varying from Middle Cambrian to Devonian are known and the limb structure is similar in all cases. There are two types of limb. First, a single pair consisting of many-jointed whip-like antennae lying one on each side of the hypostoma. Secondly, the remaining limbs which are alike except for a gradual decrease in size towards the pygidium. Each of these consists of two branches, a lower stout branch consisting of about seven joints, and an upper comb-like branch consisting of a shaft bearing a fringe of filaments.

Compared with the limbs in other aquatic arthropods, the trilobite limbs are simple. In the crustaceans, for instance, the 'head' appendages include antennae or 'feelers' and nipping pincers which hold and cut food, while other appendages may be modified for walking, swimming, breathing, filtering food or creating currents of water. By analogy, in the trilobite, the lower jointed limbs are interpreted as walking legs and the upper fringed appendages as combined swimming and breathing limbs. Specialised feeding limbs are lacking in trilobites.

The development (ontogeny) of some trilobites has been established by the comparison of specimens of the same species in different stages of growth. The earliest stage, about 1 mm long, consists of an almost circular shield with no free segments, but with the axis partly defined. In some genera the eyes are in a marginal position so that the larva may at this stage have been free swimming like larvae of many modern arthropods. Later stages show the separation of the pygidium from the cephalon, followed by the addition of free thoracic segments increasing in number at each stage until the adult complement of segments is reached. Beyond this stage, there is only an increase in size with each moult.

Skeletal form and conjectures on the mode of life of some trilobites

Calymene (fig. 112). The cephalon is the widest part of the body and is semicircular with rounded genal angles. The glabella is strongly convex, and is defined by deep axial furrows which unite in front. Incomplete glabellar furrows separate knob-like lobes on each side of the glabella.

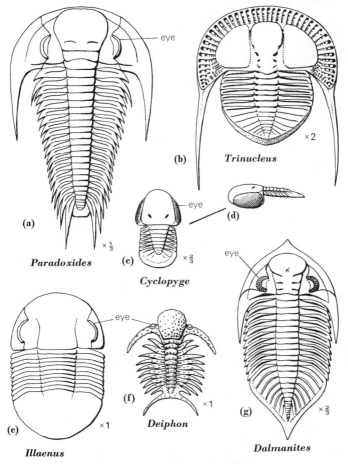

116 **Diversity of form in trilobites.**

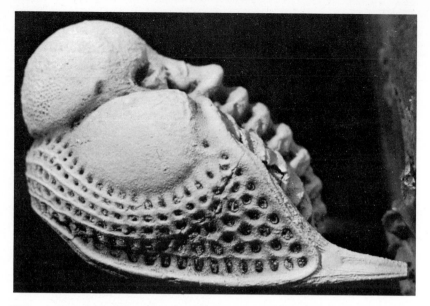

117   A trinucleid. A lateral view of *Tretaspis sortita*, Starfish Beds, Ashgill
Series, Upper Ordovician, Girvan (×6).

The eyes are small and the facial suture cuts the genal angles. The
thorax is gently tapered towards the pygidium and contains 13 segments.
The pygidium is small, with six rings in the axial region and the strongly
marked pleural furrows continue to the margin.

Silurian to Devonian.

Finely preserved specimens of *Calymene*, often in enrolled state
(fig. 112b), occur commonly in Silurian limestones (Wenlock) along
with benthonic animals like corals, brachiopods and bryozoa, and also
calcareous algae. The association suggests that the limestone was
probably laid down in clear, relatively shallow warm water. *Calymene*
itself was probably benthonic, living mainly on the sea-floor. The
evidence suggesting this includes the dorso-ventrally flattened shape of
the body, with the mouth on the under ventral side and the eyes set
high on the cheeks so that vision was restricted to a field overhead.
Limbs of typical trilobite type are known in a closely related form and
it is reasonable to conclude that *Calymene* could crawl on the sea-floor,
or swim close to it, and that it 'breathed' with fringed limbs. *Calymene*
probably moved rather slowly. Trilobites lack the streamlined torpedo
shape found in fast swimmers.

Little is known in detail about the mouth in trilobites but it was placed on the ventral surface and appears to have opened towards the posterior. Such an arrangement together with the lack of special feeding limbs and the probable lack of speed, rules out any likelihood that trilobites were active predators, or able to cope with hard-shelled organisms. The feeding habits of some modern crustaceans have provided clues to the way trilobites probably fed. They may have pushed their food forwards to the mouth with their limbs, somehow sucking or scooping it up. Possibly some forms ingested sediment for its organic content, or filtered micro-organisms with their 'gill' filaments. Other forms may have scavenged or scraped soft-bodied sessile animals or seaweed.

Paradoxides (fig. 116a) has an elongate body with a large semicircular cephalon in which the genal angles are extended backwards as long spines. The glabella is expanded towards the anterior margin and is furrowed. The facial suture is opisthoparian, and the eyes large and crescentic in shape. The thorax tapers towards the pygidium and contains about 16 to 21 segments. The pleurae end in long sharp spines directed backwards. The pygidium is very small and contains few segments.

Middle Cambrian.

Long pleural spines are a feature of *Paradoxides* and many other Cambrian trilobites. Spines have a protective function in many animals. In this instance, however, similar spines are not common in post-Cambrian trilobites. They may have been adaptive, perhaps supporting the body on soft sediment.

Trinucleus (fig. 116b) has a semicircular cephalon, larger and wider than the rest of the body. Genal spines, about twice the length of the body, project backwards. The margin of the cephalon forms a wide brim, or fringe, which has many small pits arranged in radial rows. The glabella is very convex, and is cut by two or three pairs of short deep furrows (fig. 117). The cheeks between the fringe and the glabella are convex and smooth. There are no eyes. The facial suture is largely marginal, and so is not normally seen. The thorax has six segments. The pygidium is widely triangular, and has an entire margin. Related genera include *Onnia* and *Cryptolithus* differing in small details, e.g. they lack the glabellar furrows.

Ordovician.

The significance of the fringe and long genal spines in *Trinucleus* is not understood. These features are, however, also found in a wide range of related Ordovician trilobites and must presumably have been adaptive modifications. *Trinucleus* usually occurs in fine-grained muddy rocks, and has been described as a 'mud-grubber' which used its fringe as a shovel, or alternatively as a 'snow-shoe' device to support the body on soft mud. The lack of eyes is consistent with tunnelling in mud, while typical trilobite limbs known in a related genus imply ability to crawl and swim. The pits which perforate the fringe, however, remain unexplained.

Dalmanites (fig. 116g) has a semicircular cephalon extended backwards by quite long genal spines. The glabella widens towards the anterior and is cut by short furrows. The eyes which lie close to the glabella and to the posterior margin of the cephalon, are large, and kidney-shaped with the lenses (usually distinct) on their outer curved surface. The facial suture is proparian. The thorax contains 11 segments. The pygidium is about the same size as the cephalon (isopygous), and ends in a spine.
    Silurian to Lower Devonian.
    The form and position of the large highly developed eyes suggest that *Dalmanites* was a benthonic form with a wide field of vision, and aware of movements anywhere around (including behind) it. The pygidium with its terminal spine has been compared with the tail spine of the king-crab (*Limulus*) (fig. 122). This is used by the king-crab as a lever as it scrabbles in sand for worms. The analogy cannot be complete, since the king-crab seizes its prey with 'pincers', limbs of a type unknown in trilobites.

Illaenus (fig. 116e) has a strongly convex, smooth exoskeleton of oval outline in which the trilobation is not strongly marked. It has large eyes. The facial suture is opisthoparian. There are 10 segments in the thorax, and it is isopygous.
    Ordovician.
    The most noteworthy feature in *Illaenus* is the smoothness of the exoskeleton; indeed in a related form, *Bumastus* (fig. 118), the trilobation is almost effaced. One interpretation of a smooth exoskeleton is that it was an adaptation to grubbing or even burrowing in mud, especially in forms which either lack eyes or have eyes of reduced size, as for instance *Trimerus* (fig. 119g). *Illaenus*, however, had quite large eyes and it may possibly, like various modern crustacea when at rest, have lain just covered in sediment but with the eyes protruding. An alternative mode of life is suggested by a resemblance in shape and ability to enroll between *Bumastus* and present-day marine creatures like some of the

118   A trilobite with a smooth exoskeleton, on which the axial furrows are very faint. *Bumastus barriensis*, Wenlock Limestone, Silurian ($\times 1.7$).

isopods (relatives of the wood-louse) and '*Chiton*' (a primitive mollusc with a flexible shell of overlapping plates). These animals live in the littoral and sublittoral zone, the isopods hiding in seaweed and '*Chiton*' clinging fast with its foot to the surface of a rock. In this region of turbulence, a smooth shell offering minimal resistance to wave action is an advantage; also the ability to roll up, if dislodged, reduces the likelihood of damage to the body. Trilobites which lived in a reef environment might encounter similar turbulence.

Deiphon (fig. 116f) has its cheeks reduced to form long curved spines and the eyes are set on the anterior margins, close to the very inflated glabella. The pleurae, and the pygidium also form spines.
    Silurian.
    While the spikiness of the exoskeleton must have deterred predators, it is also regarded as an adaptation to a pelagic habit, the spines increasing surface area relative to volume. The marginal location of the eyes is logical in a pelagic animal, and it has been suggested that the swollen

glabella may have contained a light substance such as fat which would have increased its buoyancy.

Cyclopyge (fig. 116c, d) has very large eyes which occupy most of the free cheeks and extend over much of the ventral surface of the cephalon; in some related forms the eyes unite ventrally.

Ordovician.

Cyclopyge had an unusually wide field of vision and it has been suggested that this was a pelagic form, possibly nocturnal in habit, rising from deeper water to feed at the surface during the night.

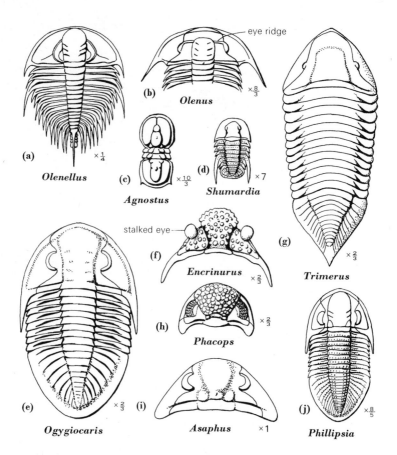

119   Trilobites.

In (a), (c), (d) the facial suture does not appear on the dorsal surface. In the remaining forms the facial suture is either opisthoparian, as in (b), (e), (i) and (j); proparian as in (f) and (h); or gonatoparian as in (g).

Additional common genera

Olenellus (fig. 119a) has a wide semicircular cephalon with genal spines. The glabella extends to the anterior border and has distinct furrows. Large crescentic eyes are joined to the front of the glabella. There is no facial suture. The anterior part of the thorax contains 14 segments complete with spinose pleurae, but the latter are absent from the remaining segments, the first of which, however, bears a long sharp spine. The pygidium is a very small plate.
Lower Cambrian.

Callavia is similar to *Olenellus* but its spines are shorter, and the long axial spine is absent.
Lower Cambrian.

Olenus (fig. 119b) is a small trilobite with an almost rectangular cephalon which is wider than the rest of the body. Sharp genal spines stick out at an angle to the body. The eyes are connected by eye ridges to the front of the glabella and the facial suture is opisthoparian. There are about 13 to 15 thoracic segments. The pygidium is small, and triangular.
Upper Cambrian.

Agnostus (fig. 119c) has a tiny (5–6 mm) isopygous exoskeleton with two thoracic segments. There are no eyes or facial suture.
Upper Cambrian. Related genera range throughout the Cambrian and into the Ordovician.

Shumardia (fig. 119d) is a tiny (5 mm) blind form without facial sutures. The thorax consists of six segments of which the fourth bears long pleural spines.
Tremadoc to Ordovician.

Asaphus (fig. 119i) is an isopygous form with eight thoracic segments. The glabella widens towards the anterior margin of the cephalon. The facial suture is opisthoparian and the free cheeks are relatively wide. The genal angles and the pleurae are rounded. The axial furrows and the segmentation are not strongly defined.
Lower Ordovician. Related forms range throughout the Ordovician; e.g. *Ogygiocaris* (fig. 119e) is common in the Lower and Middle Ordovician.

Encrinurus (fig. 119f) is an isopygous form. The cephalon is covered with tubercles. The glabella is rather inflated and is wider in front. The facial suture is proparian, and the eyes are on short stalks. The thorax

120   A Carboniferous trilobite. *Phillipsia gemmulifera*, Carboniferous
Limestone (×4).

contains 11 or 12 segments. The pygidium is sharply triangular with
many more furrows on the axis than on the pleural region.
Middle Ordovician to Silurian.

Trimerus (fig. 119g) is an isopygous form with both cephalon and
pygidium roughly triangular in shape. The glabella and axial furrows
are poorly defined, and the exoskeleton as a whole is smooth. The eyes
are small and the facial suture cuts the genal angle.
Middle Silurian to Middle Devonian.

Phacops (fig. 119h). The inflated glabella widens markedly towards the anterior margin of the cephalon and the genal angles are rounded. The eyes, large and kidney-shaped, have distinct facets and the facial suture is proparian. The thorax contains 11 segments (a characteristic of many related genera). The pygidium is small and semicircular in shape. *Phacops* is often found in an enrolled state.

Silurian to Devonian.

Phillipsia (figs. 119j, 120) has a small isopygous exoskeleton of elongated oval shape. The glabella has parallel sides and is cut by short furrows. The large crescentic eyes are set close to the glabella. The facial suture is opisthoparian. There are nine thoracic segments.

Lower Carboniferous.

Phillipsia belongs to a family, the proetids which, ranging from the Ordovician to the Permian, survived longer than any comparable group of trilobites, and during this time showed little change in the form of the exoskeleton.

Geological history of the trilobites

Trilobites appear in considerable numbers and are the dominant fossils in early Cambrian rocks. Their numbers had reached its maximum by the Middle Ordovician, and gradually declined during the Silurian and the Devonian at the end of which few families remained. They were rare in the Carboniferous, and finally disappeared in the Permian.

The earliest trilobites belong to a group typified by *Olenellus* (p. 197) which shows features common to a substantial number of Cambrian trilobites. For instance they have a large cephalon with elongate crescentic eyes, a thorax with many spinose segments, and a very small pygidium. Contrasting types like the blind isopygous agnostids (p. 197) also occur in the Lower Cambrian. Thus at their inception as fossils, the trilobites are both complex and diversified in structure, and, despite the lack of fossil evidence, this implies a relatively long existence in Pre-Cambrian time.

Trilobites occur in a variety of marine sediments, but more generally in those of relatively shallow-water origin, including reefs. Individual species may be short ranged, and so of value stratigraphically, especially in the Cambrian rocks where other types of fossils are neither common nor widespread.

In the Lower Cambrian in Britain, trilobite remains are scanty and fragmentary. The dominant genera here, *Olenellus* (fig. 119a) and

121   A Carboniferous arachnid. *Eophrynus,* **a spider-like member of an extinct order; dorsal view. Replica of a fossil preserved in an ironstone nodule, Coal Measures (×3).**

Callavia, occur in separate regions, *Olenellus* in North-west Scotland (Fucoid Beds) and *Callavia* in Shropshire (Comley Sandstone). They also occur in separate regions elsewhere. *Olenellus* is found in the Appalachian region of North America, and *Callavia* in Scandinavia and the Atlantic seaboard of North America. Many other trilobites were similarly restricted to particular regions, and a study of their distribution helps in reconstructing the geography of the period.

In the Middle Cambrian various species of *Paradoxides* (fig. 116a) have been used to establish a number of zones which can be traced over a wide area including England, Europe and North America. Forms related to *Agnostus* (fig. 119c) are also useful for correlating rocks of this age; they are abundant in argillaceous rocks and are readily distinguished. The Upper Cambrian is zoned mainly by *Olenus* (fig. 119b) and related forms. A species of *Shumardia* is characteristic of the Tremadoc.

In Ordovician times the number of trilobite genera reached its peak, and great diversity of form is shown. They include a few Cambrian forms which lingered for a while (e.g. olenids and agnostids), a number which were restricted to the Ordovician (e.g. trinucleids and asaphids), and others which ranged into the Silurian and a few beyond that. Most forms were opisthoparians, though there were also proparians and some forms with a marginal suture. *Asaphus* (fig. 119i) and *Trinucleus* (fig. 116b) represent two of the most commonly represented families. The trinucleids are useful index fossils in Britain and Scandinavia. Other characteristic forms are *Cyclopyge* (fig. 116c) and *Illaenus* (fig. 116e).

In the Silurian there was a marked increase in importance of proparian forms: *Dalmanites* (fig. 116g), *Encrinurus* (fig. 119f) and *Phacops* (fig. 119h) are examples. Other common Silurian trilobites are *Calymene* (fig. 112), *Trimerus* (fig. 119g) and *Deiphon* (fig. 116f).

Devonian trilobites were greatly reduced in numbers and variety. *Phacops* is perhaps the best known example. All those remaining at the end of the Devonian belong to one family, the proetids which, appearing first in the Ordovician, survived until the Middle Permian, the last member being little different from the earliest. Various genera like *Phillipsia* (fig. 120) may occasionally be found in Britain in the Lower Carboniferous limestones, and the latest British trilobite occurs in a marine band in the Coal Measures.

OTHER GROUPS OF ARTHROPODS WITH A FOSSIL RECORD

Arachnids are mainly terrestrial air-breathing forms represented today by spiders and scorpions. A primitive scorpion-like aquatic type, *Palaeophonus*, occurs in the Silurian in Scotland. An extinct spider-like form is shown in fig. 121.

Silurian to Recent.

Xiphosurids are represented today by the marine king-crab, *Limulus* (Tertiary to Recent, fig. 122) which has a trilobed carapace of fused segments and a long 'tail' spine; it shows a superficial resemblance to a trilobite. Occasional xiphosurids may be found in nodules in the Coal Measures (fig. 5).

Cambrian to Recent.

Eurypterids were archaic aquatic forms which were locally abundant as in some Silurian and Old Red Sandstone rocks. Some forms, like *Pterygotus* (Silurian to Devonian), which measured about 2 m, were giants amongst arthropods. *Ptergyotus* had a flattened scorpion-like

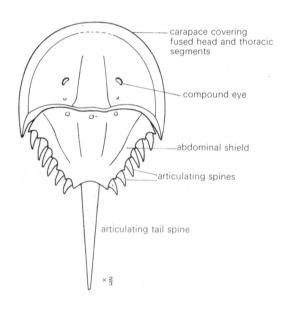

carapace covering fused head and thoracic segments

compound eye

abdominal shield

articulating spines

articulating tail spine

$\times \frac{2}{3}$

122 '*Limulus*', a Jurassic king-crab.

body and its appendages included a pair of paddle-like limbs which suggest reasonable swimming capacity, and a pair of claws which suggest carnivorous habits. A related form is shown in figs. 123, 124.

Ordovician to Permian

Insects form the largest arthropod group and are enormously varied. They are air-breathing, and typically terrestrial. There are three body regions, the head with a pair of antennae, the thorax with three pairs of legs (and typically with two pairs of wings), and the abdomen. They are the only arthropods capable of flight, and the wings, being chitinous, may be fossilised in addition to other parts of the body. Fossils, however, are rare. The earliest insects were wingless 'springtails' found in the

123   A eurypterid. *Slimonia*, Middle Silurian, Lesmahagow, Scotland ($\times \frac{1}{2}$).

124   Restoration of the eurypterid, *Slimonia* (×0·25).
(Model by Mr. F. Munro and Dr. W. D. I. Rolfe.)

125   A Jurassic dragonfly. *Protolindenia wittei*, **Upper Jurassic, Bavaria**
(×1).

126   A Cretaceous worker ant in amber. *Sphecomyrma freyi*, **Upper**
Cretaceous, New Jersey, U.S.A. This is the earliest record of a social insect.

127   A Jurassic 'prawn'. *Aeger tipularius*, **Lithographic Stone, Upper Jurassic, Solenhofen, Bavaria** (×0·6).

Rhynie Chert (Middle Old Red Sandstone). Winged insects, e.g. 'cockroaches' appeared in the Coal Measures (in non-marine beds) and the ancestors of various modern groups such as dragonflies (fig. 125) and beetles were in existence by the end of the Palaeozoic. Social forms like wasps and ants (fig. 126) are recorded from Cretaceous rocks; butterflies (Plate I) from the Cainozoic. A large variety of insects occur in amber of Oligocene age (Plate II).

The behaviour of modern insects suggests a close link between the evolution of insects and the development of flowering plants but, in the absence of fossil evidence, this remains a subject for speculation.

Devonian to Recent.

Crustaceans form a large and varied class of aquatic and terrestrial arthropods. Typically they are gill-breathing with a fixed number of segments in the body, and their appendages include two pairs of antennae, a pair of mandibles and some two-branched limbs which may be specialised for various functions.

Crabs and lobsters (decapods) are marine crustaceans which may be common as fossils in a few restricted horizons, for instance in nodules in the Gault Clay and London Clay (fig. 128). A primitive 'prawn' is shown in fig. 127.

Trias to Recent.

128   An Eocene crab. _Rachiosoma_, London Clay (×1·3). View of part of the dorsal surface and of the claws.

Ostracods (fig. 143m) are tiny crustaceans with the body enclosed by a calcareous, two-valved carapace of oval shape. They occur in marine, brackish-water and fresh-water deposits, often in great abundance, and may be of stratigraphic value. They are common in some beds in the Purbeck and Wealden.
Ordovician to Recent.

Waterfleas (branchiopods) are tiny aquatic crustaceans with the body enclosed by a two-valved chitinous carapace. One form occurs today in temporary (playa) lakes in South Africa, and closely similar fossil forms, _Euestheria_ (fig. 143n) for instance, occur in some horizons in the Coal Measures and in the Trias.
Devonian to Recent.

TECHNICAL TERMS

ANTENNAE many-jointed whip-like appendages. Trilobites have one pair attached one on each side of the hypostoma (fig. 115).

APODEME a paired process on the inner surface of the dorsal exoskeleton near the axial furrow to which the limb muscles may have been attached (fig. 115).

AXIS central region of the trilobite defined on each side by a longitudinal furrow (fig. 112a).

CARAPACE dorsal exoskeleton covering part of the body in some arthropods, e.g. crustaceans.

CEPHALON the anterior part of the dorsal exoskeleton which covered the head region (fig. 112a).

CHEEK the area of the cephalon on each side of the glabella. The cheeks may be separated by the facial suture into the fixed and free cheeks (fig. 112a).

CRANIDIUM the central region of the cephalon comprising the glabella and fixed cheeks (fig. 112e).

DOUBLURE the margin of the dorsal exoskeleton which is reflexed on to the ventral side (fig. 112e).

ECDYSIS moulting of the exoskeleton in arthropods.

EYE RIDGE a fine ridge between the eye and the anterior part of the glabella in some trilobites, e.g. *Olenus* (fig. 119b).

FACIAL SUTURE a fine suture which separates the free cheek from the cranidium on each side of the head (fig. 112a); it runs from the posterior or lateral margins of the cephalon across the cheeks and along or under the anterior margin.

GENAL ANGLE the angle between the posterior and lateral margins of the cephalon (fig. 112a). It may be rounded or produced into a genal spine (fig. 116a).

GLABELLA the arched axial region of the cephalon. It is separated from the cheeks by the axial furrows and may be cut by transverse glabellar furrows (fig. 112a).

HYPOSTOMA a small plate lying in front of the mouth region on the ventral side of the cephalon (fig. 112e).

LIMBS a pair of appendages attached to each segment of the body on the ventral side; they comprise (i) one pair of single-branched antennae, and (ii) many pairs of similar two-branched limbs, each limb consisting of an outer gill-bearing branch and an inner walking leg made up of about seven joints (fig. 112d).

OCCIPITAL FURROW a transverse groove across the glabella which separates the posterior segment, the occipital ring, from the rest of the glabella (fig. 112a).

OPISTHOPARIAN a facial suture which cuts the posterior margin of the cephalon (fig. 116a).

PALPEBRAL LOBE a raised area on the fixed cheek against which the eye abuts.

PLEURAE the lateral parts of a thoracic segment (fig. 112a).

PLEURAL FURROWS furrows, usually oblique, on the pleurae.

PROPARIAN a facial suture which cuts the lateral margin of the cephalon (fig. 116g).

PYGIDIUM the tail-shield of the exoskeleton covering the fused posterior segments of the body (fig. 112a).

THORAX the separate, freely articulating segments interposed between the cephalon and the pygidium (fig. 112a). The number of segments varies in different genera.

14

Graptolithina

Graptolites are the remains of extinct colonial marine organisms which are confined to Palaeozoic rocks. They form a class which has been referred to various groups but is now generally included, along with the pterobranchs, in a subphylum of the Hemichordata.

The graptolites secreted a proteinaceous skeleton, the RHABDOSOME, which originated in a tiny conical cup, the SICULA. From this grew one or more long slender branches, called STIPES, which were either free or were linked by short cross-connections, and each of which was made up of a linear series of short overlapping tubes, the THECAE. Each of these housed an individual member (ZOOID) of the colony. The rhabdosome shows bilateral symmetry.

Two orders are important as fossils, the Graptoloidea, or true graptolites, and the Dendroidea or dendroid graptolites. The Graptoloidea have a particular geological importance as zonal fossils in Ordovician and Silurian rocks in which they occur, often in great numbers, in dark shales which for the most part lack other types of fossils. They have two characteristics which make them particularly suitable for use as zonal fossils. Firstly, the various species and genera have comparatively short ranges, following one another in relatively rapid succession; and secondly, having apparently followed a planktonic mode of life, they are widely distributed throughout the world. The Dendroidea, by contrast, have a restricted occurrence and their main interest as fossils centres on the clues they provide to relationships between graptolites and other animal phyla, especially the Hemichordata. The dendroid graphites are considered here only briefly.

ORDER GRAPTOLOIDEA

In the Graptoloidea the rhabdosome consisted of one or a small number of stipes each comprising a series of similar thecae, and growing from the sicula in a definite pattern.

PRESERVATION. Graptolites are most commonly found in black shales as carbonised impressions, or whitish films resembling tiny fret-saw blades. More rarely, uncrushed specimens are found in limestone (Plate III) or chert. These can be freed by dissolving the matrix with a dilute chemical (e.g. acetic acid for limestone, hydrofluoric acid for chert) to which the fossil material is resistant. Often the substance of the skeleton in such specimens is only slightly carbonised and, if embedded in paraffin wax, can be cut in serial sections with a microtome. From such material, the morphology and development of a number of grapto-lites is known in great detail.

Morphology

The graptolite skeleton, the PERIDERM, is a scleroproteinous material (p. 8) which in transmitted light is a rich brown colour (Plate III). It has two layers and the inner one shows growth bands in the form of alternating half-rings which dovetail into each other with a zigzag suture (fig. 129b).

RHABDOSOME. Stages in the development of the sicula and rhabdosome in a number of different genera are known in some detail. The SICULA was secreted by the first member (a zooid) of the colony and the THECAE were formed by the subsequent zooids by a process of budding which follows a distinct and characteristic pattern in each of several families of graptolites.

The SICULA (fig. 129c) is a conical structure about 1·5 mm in length. Its apex is extended as a thread-like tube, the NEMA, and its wide end, the APERTURE, is open. A spine, the VIRGELLA projects from one side of the aperture.

The thecae (fig. 129a) are basically short tubes arranged in an over-lapping series along the stipe. Each opens internally in a passage, the COMMON CANAL, shared with the other thecae; the external opening is the APERTURE.

There is a considerable range in shape of thecae and this is of value in the recognition of species and genera. Thus the theca may be straight, i.e. SIMPLE (fig. 130a) or may show S-shaped or sigmoidal curvature (fig. 130d). In some forms the apertural end is bent over, i.e. HOOKED (fig. 130b), or may be twisted to one side. The thecae may overlap closely, or may be widely separated, ISOLATE (fig. 130e). The aperture may be circular in shape or constricted. Spines occur on some forms.

The RHABDOSOME may consist of one, two, four or many STIPES. The angle at which the stipes diverge from the sicula is fairly constant in a given species, and in most forms with more than one stipe the branching of the stipes is symmetrical (dichotomous). The rhabdosome is described as PENDENT (fig. 130h) if the stipes grow downwards from the sicula with the thecae facing inwards. It is SCANDENT (fig. 130d) if the stipes grow upwards from the sicula with the thecae facing outwards. Other positions of the stipes relative to the sicula include HORIZONTAL (fig. 130g) and RECLINED (fig. 130f). The stipes may consist of a single row of thecae, UNISERIAL (fig. 130a) or of two rows of thecae growing back to back, BISERIAL (fig. 130d). Biserial forms are also scandent.

The nema (fig. 129c) is typically present, although in some graptolites

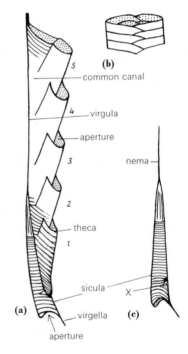

Monograptus

129   Morphology of the graptoloids.

a, the proximal end of the stipe of a scandent, uniserial rhabdosome (the thecae, of simple type, are numbered 1 to 5). b, a part of the inner layer of the periderm, showing growth lines arranged in alternating half-rings which meet along a zigzag suture. c, sicula; X is the point of origin of the first theca.

it may be very short. In scandent forms the nema is referred to as the VIRGULA (fig. 130d) and it is incorporated in the rhabdosome.

Some common graptoloids

Monograptus (figs. 130a–c and 131). The rhabdosome consists of one stipe, either straight or curved; it is uniserial and scandent, supported by the virgula which extends distally from the sicula. The aperture of the sicula faces downwards, and the first theca grows upwards. The thecae show a considerable diversity of shape; they may be simple, show slight or marked sigmoidal curvature, may be isolate or hooked. In some species the distal (last-formed) thecae differ in shape from the proximal (first formed) thecae.
Silurian

Rastrites (fig. 130e) resembles *Monograptus*; the rhabdosome is curved, and the thecae are long, widely isolate and hooked distally.
Lower Silurian (Middle and Upper Llandovery).

Cyrtograptus (fig. 130i) is a *Monograptus* with a spirally coiled rhabdosome to which lateral 'branches' (cladia) are attached.
Middle Silurian.

Diplograptus (fig. 132d) is biserial and scandent. The thecae are straight or show slight sigmoidal curvature and they form a closely overlapping series with the apertures facing outwards.
Middle Ordovician to Lower Silurian (Llanvirn to Llandovery).

Climacograptus (fig. 132e, f, Plate III) resembles *Diplograptus* but the thecae are strongly sigmoidal and the apertures face upwards.
Ordovician to Lower Silurian.

Dicellograptus (fig. 132b, c) is uniserial with two reclined stipes. The thecae show pronounced sigmoidal curvature.
Middle and Upper Ordovician.

Dicranograptus (fig. 132a) resembles *Dicellograptus* distally but the early part of the rhabdosome is biserial and scandent.
Middle Ordovician.

Nemagraptus (fig. 132g) has two slender stipes which diverge at about 180° from the sicula and curve to form an S-shaped rhabdosome; lateral stipes arise at intervals from the convex sides of the two main stipes. The long slender thecae show sigmoidal curvature.
Middle Ordovician.

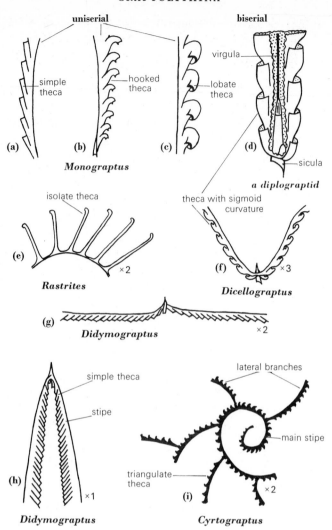

130   Variations in the form of the thecae and rhabdosome in graptoloids.
The rhabdosome is uniserial in each case except in (d), where it is biserial; it is scandent in
(a)–(e) and (i), reclined in (f), horizontal in (g) and pendent in (h). In (d), part of the rhabdosome
is broken, exposing the sicula and virgula.

Didymograptus (fig. 130g, h), has a uniserial rhabdosome with two stipes diverging at about 180° or less from the sicula. In most species the stipes are either horizontal or pendent. The thecae are simple.

Lower and Middle Ordovician, mainly Arenig and Llanvirn.

Tetragraptus (fig. 132i, k, l), has four uniserial stipes which may be pendent,

**131 *Monograptus*; an uncrushed specimen extracted from limestone.
Lateral view showing the sicula, the first three thecae and growth lines.
Monograptus chimaera, Lower Ludlow, Silurian (×60).**
(Photographed in transmitted light.)

horizontal or reclined. The thecae are simple and closely overlapping.
Lower Ordovician (Arenig).

Phyllograptus (fig. 132h) is a scandent *Tetragraptus*.
Lower Ordovician.

Dichograptus (figs. 132j and 133) is uniserial; there are eight stipes and branching is dichotomous. The thecae are simple, and overlap closely. Lower Ordovician.

Mode of life of the graptoloids

Since the graptoloids are extinct and have no living close relatives, evidence concerning their mode of life is necessarily circumstantial. The evidence is partly sedimentary and partly biological.

The graptoloids sometimes occur together with trilobites, brachiopods and other marine organisms and were themselves, therefore, marine. More usually, however, they occur in great numbers in dark shales which contain few or no other fossils. Such shales were formed in quiet, relatively deep water. Their dark colour is due to organic matter derived from planktonic plants and animals living in the photic zone, whose remains sank and accumulated on the sea-bed. Such organic matter is a source of food for many animals and is also susceptible to destruction by oxidation. Its persistence, therefore, is an indication that few feeding organisms were present and that the oxygen concentration at the sea-bottom was low. The shales also contain the ferrous sulphide, pyrites, which forms in reducing conditions either within the sediment or, if the bottom waters are completely deficient in oxygen, on the sea-floor itself. In the latter case, the production of H_2S may make the water toxic to most forms of life. This environment, therefore, is one which is un-favourable or even hostile to benthonic organisms. On the other hand, it is an environment favourable to the preservation of floating organisms which might sink to the bottom after death, since scavengers would be few or absent. It is most likely, therefore, that the graptoloids were planktonic organisms, possibly attached to seaweed, living in the well-oxygenated surface waters and widely distributed by marine currents. Their remains accumulated in quantity, however, only in the dark mud environment. Elsewhere, turbulent conditions and the presence of scavengers would ensure the destruction of the majority of the dead bodies.

Observed facts about the wide geographic distribution of the grapto-loids, and about the nature of the rhabdosome are in accord with the conclusions reached above. The evidence that they were not benthonic creatures lies in the lack of any rooting device in true graptolites by which they might have been anchored to the sea-bed, and also in the bilateral symmetry of the rhabdosome which is not in general a charac-teristic of sessile organisms. Two further features which are best explained with reference to a planktonic mode of life are firstly, the

nema, which is generally accepted as a means of attaching the rhabdosome to floating debris or seaweed; and secondly, the association of
certain biserial graptoloids in groups around a disc-like object as if in
life it floated overhead and the graptoloids hung down from it.

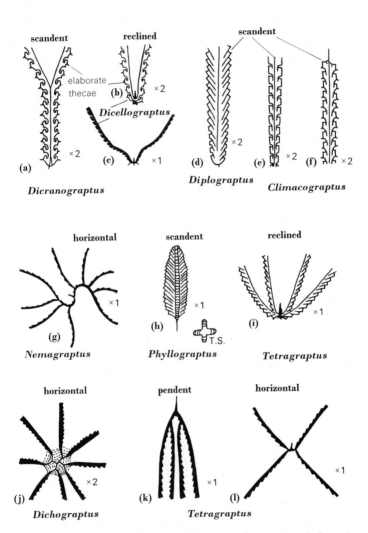

**132 Variations in the number and direction of growth of the stipes in
graptoloids.**

The rhabdosome is uniserial in each case except in (a), where it is unibiserial, and (d)–(f),
where it is biserial, and (h), where it consists of four stipes. Note also the differences in the form
of the thecae.

133   A many-branched graptoloid. *Dichograptus octobrachiatus*, Ordovician, Victoria, Australia (×1·5).

Geological history of the graptoloids

True graptolites appeared in the early Ordovician and proliferated rapidly to become a widespread and dominant part of the marine fauna throughout the Ordovician and until about the middle of the Silurian, after which numbers began to decline. They finally disappeared early in the Devonian (Gedinnian).

The earliest graptoloids are dichograptids, mainly uniserial forms, some with many stipes, e.g. *Dichograptus* (eight, fig. 132j), others with few stipes, e.g. *Tetragraptus* (four, fig. 132i, k, l) and *Didymograptus* (two,

fig. 130g,h). All have thecae of simple type. In the main they are
horizontal and pendent types but a number are scandent. Species of
Didymograptus are the zonal indices in the Arenigian and Llanvirnian,
horizontal types characterising the Arenigian, and pendent forms the
Llanvirnian.

The dichograptids are succeeded in the Caradocian and Ashgillian
by forms with two uniserial reclined stipes and slender thecae showing

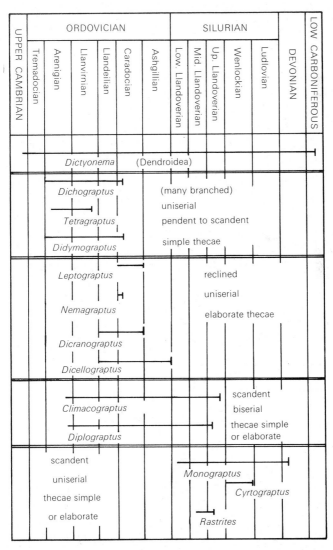

134 Summary of the geological distribution of the graptolites.

sigmoidal curvature. One form, *Nemagraptus gracilis* (fig. 132g) may be cited here to illustrate the value of certain species in world-wide correlation; this species has a restricted vertical range at the base of the Caradocian, but it is found in Britain, Europe, North America, Asia and Australia. The characteristic Ashgillian form is *Dicellograptus* (fig. 132b, c), some species of which show very pronounced sigmoidal curvature of the thecae. In addition, biserial scandent graptoloids are present in large numbers throughout the Caradocian and Ashgillian, e.g. *Diplograptus* (fig. 132) and *Climacograptus* (fig. 132e, f), and these continue into the Silurian.

Uniserial scandent graptoloids make their appearance in the Silurian. They are referred for the most part to *Monograptus* (fig. 130a–c) and numerous species have been distinguished mainly on the basis of the thecal shape. They can be separated broadly into groups which succeed each other in the Silurian. The early part of the Llandovery is characterised by monograptids with quite simple thecae, associated with many diplograptids. In the Middle Llandovery the monograptids have more elaborate thecae and are associated with dwindling numbers of diplograptids. In the Upper Llandovery monograptids occur alone, and have isolate or hooked thecae. *Rastrites* (fig. 130e) is a characteristic example. Wenlock monograptids have mainly hooked thecae (fig. 130b). *Cyrtograptus* (fig. 130i) is restricted to this horizon. In the Ludlow most monograptids have relatively simple thecae; the fauna can, as a whole, readily be separated from that of the Llandovery by the absence of biserial graptoloids.

ORDER DENDROIDEA

The rhabdosome of dendroid graptolites differs from that of true graptolites in that it consists of many stipes, each made up of three distinct kinds of thecae which form a regular alternating series; the stipes may be connected by transverse bars.

Morphology

The dendroid graptolites are usually preserved as a flattened carbonised film in shale. Relatively uncrushed material from siliceous nodules has provided information about the finer details of their structure.

The rhabdosome (fig. 135a) consists of many branching uniserial stipes which grow in a pendent or horizontal direction and which may be connected by cross-bars. The thecae face downwards from the sicula, and in compressed fossils are often concealed.

The dendroids contrast with true graptolites in possessing three kinds of theca which arise in regularly alternating triads from an internal scleroproteinous structure, the STOLON (fig. 135c). The most conspicuous theca is a simple tubular structure like the true graptolite theca (fig. 135b). The second type of theca (BITHECA) is similar in shape but smaller in size (fig. 135b). The remaining type of theca is essentially an internal tube from which the next triad of thecae arises. The stolon extends along the axis of the stipe, and from it small branches run to the base of each of the three thecae of the successive generations of thecae along the stipe (fig. 135c).

A common dendroid graptolite

Dictyonema (fig. 135a) is a common dendroid graptolite with a conical net-like rhabdosome; the compressed fossil has a triangular outline. The stipes branch symmetrically and are joined at intervals by transverse bars. A nema may be present.

Upper Cambrian to Lower Carboniferous.

Mode of life of dendroid graptolites

Most dendroids are thought to have been sessile animals, attached by the apex of the rhabdosome, which in these cases was thickened and root-like. Others have an encrusting habit. They are associated most frequently with shallow-water benthonic animals, and their distribution is sporadic. Some forms, however, had a nema and may have floated suspended from seaweed; these have a more widespread distribution.

Geological history of the dendroid graptolites

The dendroids appeared in the Upper Cambrian and were well established by late Cambrian. They range upwards to the Lower Carboniferous and show very little change in structure. Their occurrence is infrequent. *Dictyonema* (fig. 135a), a typical example, is a long-ranged and conservative genus.

THE CLASSIFICATION OF THE GRAPTOLITES

Graptolites have been variously classified in the past, for instance with the Bryozoa and with the hydroids (Coelenterata). Uncertainty about their affinities is not surprising, for their normal mode of preservation reveals little of their detailed structure. Examination of uncrushed dendroids from siliceous nodules, first described in 1938, has shown that they may be most closely related to the Pterobranchia, a class in the phylum Hemichordata. The pterobranchs are sessile colonial organisms which secrete a scleroproteinous exoskeleton (fig. 135d). This shows transverse growth bands and has a branching structure. Each branch consists of a series of slender tubes occupied by zooids. These show bilateral symmetry and they are attached to an internal structure, the stolon. They feed on micro-organisms filtered by ciliated tentacles. Primitive pterobranchs are found in rocks of Tremadoc age; the next known fossil records are in Cretaceous and Tertiary rocks. A modern pterobranch, *Rhabdopleura* (fig. 135d) lives in the North Atlantic and the

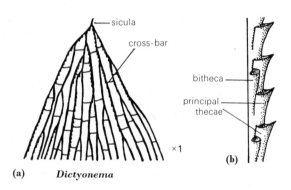

(a)   *Dictyonema*

(b)

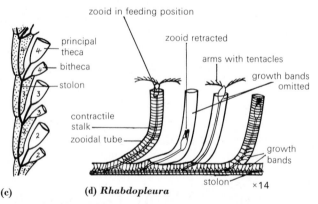

(c)   (d) *Rhabdopleura*

135   Morphology of a dendroid graptolite.

a, part of a rhabdosome. b, part of a stipe showing the arrangement of the larger and smaller thecae. c, generalised diagram to show the development of the thecae, in triads, from the internal tubular theca. d, part of a living pterobranch.

North Sea. The skeletons of graptolites and pterobranchs are similar in the following respects:

(i) They consist of scleroproteinous material.
(ii) They show transverse growth bands arranged in half-rings (shown diagrammatically in fig. 129b).
(iii) The tubes (thecae) occupied by the zooids show bilateral symmetry.
(iv) An internal structure, the stolon, is present in dendroid graptolites and pterobranchs.

A SUMMARY OF THE CHANGES SHOWN BY THE GRAPTOLITES

The dendroids are the first graptolites to appear, and since some of the early graptolites are transitional in their characters between the dendroid and the true graptolites the origin of the true graptolites is undisputed. The changes which occurred when the latter separated from the dendroids included the loss of the small thecae, the bithecae, and the loss of the cross-bars between the stipes; they also lacked the internal stolon.

In detail the evolution of the true graptolites has actually a quite complex pattern. However, the broad faunal succession which can be traced in the rocks provides a simplified picture with a few well defined lines of change. In the first instance early graptoloids have many stipes; later ones have few, or two stipes and the latest have only one stipe. Then, while in early forms the thecae are mainly simple, in later forms the thecae may be quite elaborate. The direction of growth of the stipes relative to the nema was restricted to a small number of positions. The early graptoloids were mainly pendent or horizontal forms, though scandent biserial forms co-existed with them. The latter persisted when reclined forms replaced the pendent and horizontal forms and survived for a time alongside the scandent uniserial *Monograptus* which appeared abruptly when the reclined graptoloids disappeared.

TECHNICAL TERMS

BISERIAL describes a rhabdosome in which two rows of thecae are united back to back and grew in an upwards direction from the sicula.
COMMON CANAL the cavity along the axis of the stipe which contains the stolon, and into which the thecae open.

DENDROID refers to members of the Dendroidea; it also describes an irregular bushy branching habit.

DISTAL END last-formed part of the skeleton.

DORSAL side of the rhabdosome opposite to the apertures of the thecae.

EXTENSIFORM refers to horizontally growing stipes.

ISOLATE describes the wide spacing of adjacent thecae along a stipe.

NEMA the thread-like extension of the apex of the sicula.

PENDENT refers to a rhabdosome in which the stipes hang down from the sicula.

PROXIMAL the first-formed part of the rhabdosome.

RHABDOSOME the skeleton of the graptolite.

SCANDENT describes stipe(s) growing erect along the virgula.

SICULA skeleton of first member (zooid) of the colony.

SIGMOIDAL S-shaped bending of the theca.

SIMPLE refers to a theca with straight tubular shape.

STIPE one branch of the rhabdosome made up of a linear series of thecae.

STOLON an internal thread-like structure found in dendroid graptolites, from which the zooids originated.

THECA the tubular structure in which a zooid was housed.

UNISERIAL describes stipes consisting of a single series of thecae.

VENTRAL the side of the rhabdosome bearing the apertures of the thecae.

VIRGELLA spine projecting from the margin of the aperture of the sicula.

VIRGULA name given to the nema when it is incorporated in the rhabdosome of scandent forms.

15

Porifera, Bryozoa and Annelida

PORIFERA

The Porifera, more usually called SPONGES, are sessile aquatic animals with the simplest structure found in the many-celled animals. They are mainly marine, and live in all depths of water. Sponges are often encrusting, and may be of irregular shape, but in many the body has a definite form, e.g. bowl- or vase-shaped with approximate radial symmetry. The skeleton is internal, and may be *organic* (spongin as in bath sponges), *siliceous* (opaline silica) or *calcareous* (calcite).

In its simplest form, the body is sac-shaped (fig. 136a). The wall is perforated by small pores leading by a system of canals to a central cavity which is lined by special cells, flagellate collar cells. A current of water, initiated by the beating of the flagellae, is drawn through the pores into the central cavity where the collar cells remove food (microorganisms) and oxygen. The water is passed out by one or more large openings (OSCULUM, -I), on the upper surface (fig. 136a).

The body wall consists of an inner and outer layer sandwiching a jelly-like substance with cells which secrete the units (SPICULES) of which the skeleton is composed. The spicules vary in shape but are usually symmetrical (fig. 136f–m). They may remain discrete and thus separate on the death of the animal, or they may be united to form a rigid structure, in which case forms with calcareous or siliceous spicules may be preserved entire.

Some common fossils

Siphonia (fig. 136c, d). The skeleton is tulip-shaped and was attached by a stalk. Minute surface pores lead to a system of canals which traverse the thick wall, some in a radial direction and others parallel to the outer surface (fig. 136d). Siliceous.

Middle Cretaceous to Tertiary.

Ventriculites (fig. 136e). The skeleton is roughly bowl-shaped, varying from deep to shallow, and was attached by 'rootlets'. The pores lead to simple radial canals. Siliceous.

Middle and Upper Cretaceous.

Raphidonema (fig. 137). The skeleton is vase- or funnel-shaped with a rough outer surface and a somewhat indistinct canal system. Calcareous.

Trias to Cretaceous.

Mode of life

Sponges are gregarious, living for the most part in relatively clear water. They are found at all depths from the littoral zone down to abyssal depths. Modern calcareous forms are usually restricted to coastal waters less than 100 m deep while siliceous sponges range into deeper water to about 300 m though a few of these are characteristically found in cold water below 1000 m.

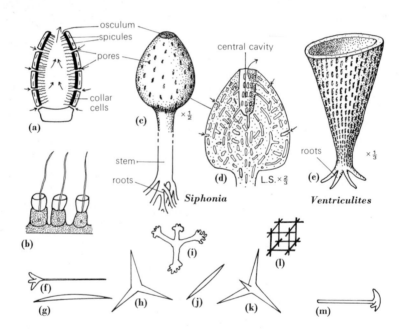

136 Morphology of the sponges.

a, longitudinal section through a simple type of sponge to illustrate the general structure of the body; b, enlarged view of collar cells. c–d, tulip-shaped sponge (d, longitudinal section showing the system of radial and vertical canals; the latter lead into the central cavity). e, vase-shaped sponge. f–m, various types of spicules. (Arrows indicate the incurrent and excurrent flow of water.)

137 A calcareous sponge. *Raphidonema*, Lower Greensand, Lower
Cretaceous, Faringdon, Berks.

Geological history

Range: Cambrian to Recent.

Lower Palaeozoic sponges had siliceous skeletons; forms with calcareous skeletons appeared in the Devonian. Isolated spicules of sponges may be common, especially in cherts and silicified limestones as, for instance, in the Lower Carboniferous. Otherwise, remains of sponges are only locally abundant occurring mainly in rocks of shallow-water origin such as sandstones and calcareous rocks. In Britain they are most common in the Cretaceous, e.g. in the Faringdon Gravels and also in the Chalk where they are often enclosed in flint nodules (fig. 138).

BRYOZOA

The Bryozoa are tiny colonial aquatic organisms which typically have a protective skeleton. They are sessile and most live in the sea where they occur from tide level down to great depths. They include the 'sea mats' which encrust seaweeds, and the freely branching 'sea-mosses' which are common on many modern beaches, and also an abundance of fossil forms.

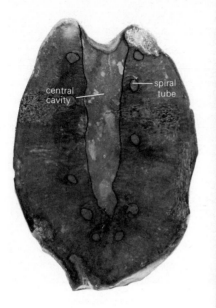
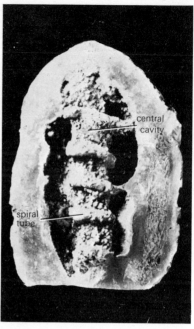

138   An association of a sponge (*Siphonia*) and a worm in flint nodules from the Chalk, Upper Cretaceous.

The worm lived within the sponge body in a spiral tube which opened to the surface at the margin of the osculum. Left, a polished median section of a sponge impregnated with silica; the circles are sections of the mould of the worm tube. Right, a mould of the central cavity of a sponge and of the spirally coiled worm tube (× 0·6).

The bryozoan colony, or ZOARIUM, contains many individual animals, or ZOOIDS. The skeleton may be horny but in the forms most commonly occurring as fossils it is calcareous. The zoarium shows a considerable range in form and size. It may be an encrusting, stick-like or massive structure, or a delicate branching network which usually measures only a few centimetres. It consists of tubular or box-like chambers, the ZOOECIA, which are open at one end, the APERTURE, and which are rarely more than a millimetre in size (fig. 139a).

The zooecium is secreted by the body wall, and the entire soft body can be retracted within its confines. The body is relatively complex, somewhat similar to that of the brachiopods. The mouth is surrounded by a ring of ciliated tentacles, and leads into a U-shaped digestive tract ending in the anus which opens near the mouth. During feeding, the tentacles extend through the aperture and filter micro-organisms (e.g. diatoms) from the sea-water.

A common fossil bryozoan

Fenestella (figs. 139b, 140). The zoarium is fan- or funnel-shaped consisting of a network of branches which are united at intervals by cross-bars. The zooecia are short tubes bound together in calcareous tissue and with the apertures forming two rows along the front side of the zoarium.

Ordovician to Permian.

Mode of life

Living Bryozoa adhere to seaweeds, shells or stones. They are found mainly in clear water but will tolerate muddy conditions. They are common in coral reefs. Fossil Bryozoa occur most abundantly in impure

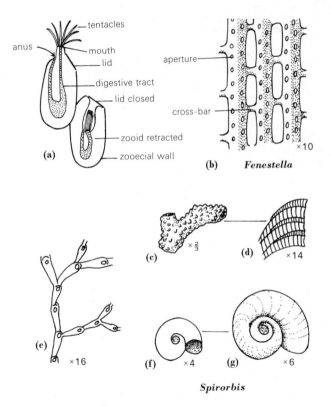

139 Bryozoans and an annelid.

a–e, bryozoans; a, two zooecia showing a simplified arrangement of the soft parts. b, a fragment of a zoarium of *Fenestella*. c, d, a slender branching (stick) bryozoan; c, a fragment of the zoarium and d, a section showing the tubular zooecia with partitions. e, part of an encrusting linear series of zooecia. f, g, an annelid tube.

140 *Fenestella*, **Carboniferous Limestone** ($\times 2 \cdot 6$).

limestones and reef limestones; they may also be found in shales but rarely in sandstones. The massive forms are often well preserved but the more delicate lacy zoaria tend to be fragmentary.

Geological history of the bryozoans

The Bryozoa range from the early Ordovician to the present day. They are abundant at many horizons and remain a flourishing group.

A variety of bryozoans are short ranged and widely distributed and are accordingly valuable for purposes of correlation. The predominant Lower Palaeozoic forms had a massive or stem-like zoarium and tubular zooecia. They played an important part (along with tabulate corals) in reef formation. They are common, for instance, in the Wenlock Limestone. The commonest type of Bryozoa in the Upper Palaeozoic had a lacy net-like or branching zoarium in which the apertures of the zooecia were recessed. *Fenestella* (fig. 140) is a long-ranged example which is

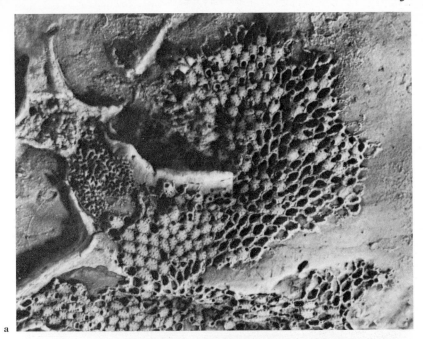

a

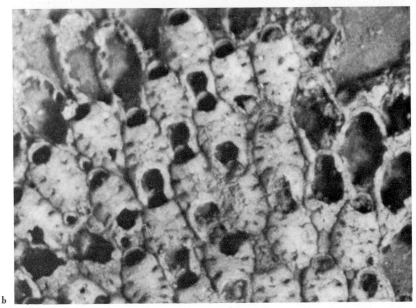

b

141 A Cretaceous bryozoan, with box-like zooecia, encrusting an oyster shell. *Leptocheilopora*, Upper Cretaceous, Sweden.

a, part of a zoarium (×3·3). b, detail of the zooecia (×40).

142   *Canadia*, an annelid from the Burgess shales, Middle Cambrian, British Columbia (×10).

Part of a free-moving marine worm showing (i), a series of segments, each with a pair of 'legs' bearing bundles of bristles and (ii), the straight gut lying along the centre of the body. (Photograph taken in ultra-violet light of a fossil in the Collection of the Geological Survey of Canada.)

common in the Magnesian Limestone of the Permian. These two groups did not survive the Palaeozoic, but other types less conspicuous in the Palaeozoic ranged on through the Mesozoic to the present day.

Bryozoa with rounded box-like zooecia and with a lid to close the aperture appeared in the Jurassic; they were encrusting or formed delicate branching stems or fronds. They were abundant in the Cretaceous, especially in the Chalk (fig. 141), and are also the dominant Bryozoa of the Tertiary and present day. The Coralline Crag is particularly rich in fragments of Bryozoa.

ANNELIDA

Annelids are segmented worms which lack a skeleton, although a few forms secrete a calcareous tube, and some have horny jaws. They are widespread and abundant today, marine forms playing an important geological role in scavenging and churning up soft sediment on the sea-floor which they ingest for its organic content. Their fossil record is scanty, but long. Their tracks and burrows date from Pre-Cambrian times. Their record includes finely preserved impressions of the soft body occurring in the Burgess Shales of Middle Cambrian age in British Columbia (fig. 142), and calcareous tubes of forms like the spirally coiled *Spirorbis* (Ordovician to Recent, fig. 139g), which may often be found on shells.

Range: Pre-Cambrian to Recent.

16

Microfossils

Microfossils are the remains of diverse microscopic animals and plants, whose study requires the use of a microscope. Animal remains include Foraminifera, radiolarians (both protozoans), ostracods (p. 207), sponge spicules (p. 225) and conodonts; Bryozoa (p. 227) and graptolites might also be regarded as microfossils since although the entire colonies are macroscopic, the finer details of individual zooecia and thecae (important for identification) can only be seen when magnified. Plant remains, referred to in later chapters, include diatoms and coccoliths, the latter so minute that only the ultra-high magnification provided by the electron microscope can reveal their detail (figs. 189, 190).

PROTOZOA

The Protozoa include a diversity of one-celled aquatic organisms, most of which are less than one millimetre in diameter. A shell is secreted by members of two groups, the Foraminifera and the Radiolaria, which are common as fossils.

Foraminifera

The foraminiferan body consists of a blob of protoplasm which carries out all the bodily functions of respiration, feeding, excretion and reproduction which in the many-celled animals are performed by separate organs. The protoplasm lies mainly within the shell, or TEST, but some extends through one or more openings, APERTURES, to enswathe it, and form thread-like extensions, the PSEUDOPODIA, which are used for movement, and to capture other micro-organisms like diatoms for food.

The test is calcareous, horny or agglutinated, i.e. made up of foreign particles like sand or shell fragments cemented by an organic matrix. It may consist of a single chamber (fig. 143c) or more usually, of several communicating chambers which are arranged spirally (fig. 143g), or in a

row or zigzag line. The size range is from a few hundredths of a millimetre to a few centimetres.

Foraminifera are mainly marine but a few forms live in brackish
water and fresh water. They are enormously abundant in the sea,
occurring at all depths from the shore-line to the abyssal zone, and in all
latitudes from the poles to the equator. In parts of the ocean where little
detritus accumulates, the remains of calcareous Foraminifera may build
up extensive deposits, or oozes. Fossil forms, too, may constitute much
of the bulk of certain limestones, as in, for example, the nummulitic
limestones from which the Egyptian Pyramids were in part built.

Mode of life

Most Foraminifera are benthonic, but a number, e.g. *Globigerina* are
planktonic. The benthonic forms may be sessile or vagrant, the latter

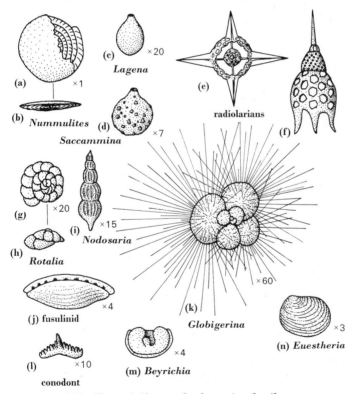

143 Foraminifera and other microfossils.

a–d and g–k, foraminifera. e and f, radiolarians. l, a conodont. m, an ostracod. n, a
branchiopod.

crawling sluggishly by means of their pseudopodia. The majority occur in the neritic zone, and here the test is often a flattened disc shape or is coiled in a flat top-like spiral. Planktonic forms are most abundant at shallow depths, between about 6 and 30 m; they are dispersed by currents, and they often have a globular calcareous test with fragile spines (fig. 143k).

Many species are restricted in their occurrence to a narrow range of physical conditions, being especially sensitive to temperature and salinity, and in the case of benthonic forms to the nature of the sea-bed (e.g. whether sandy or muddy) and to turbulence. Fossil forms, too, show an environmental preference, though this may be obscured by accidental transport (by currents) after death. Broadly speaking, calcareous Foraminifera are commonest in warm regions, and since temperature falls with depth of water, in surface waters. Larger forms, too, are mainly restricted to warm areas. Agglutinated forms are more characteristic of colder waters. Some common examples are:

Saccammina (fig. 143d). The test is agglutinated and consists of a globular chamber with a short neck. Benthonic.
   Silurian to Recent.

Lagena (fig. 143c). The test is calcareous, and consists of a single flask-shaped chamber. Benthonic.
   Jurassic to Recent.

Nummulites (figs. 143a, b and 144). The test is calcareous. It is a flattened disc shape made up of many whorls coiled in a plane spiral, each whorl enveloping the preceding one. The whorls are divided by septa into chambers, each connected with its neighbour by a slit-like opening. Benthonic.
   Lower Tertiary.

Rotalia (fig. 143g, h). The test is calcareous. The chambers are coiled in a low spiral with all the whorls showing on one side, but only the last whorl on the other side. Benthonic.
   Upper Cretaceous to Recent.

Globigerina (fig. 143k). The test is made of calcite and is perforated by many fine pores. It consists of several spherical chambers, coiled in an irregular spiral. Planktonic.
   Tertiary to Recent. Related forms occur in the Upper Cretaceous.

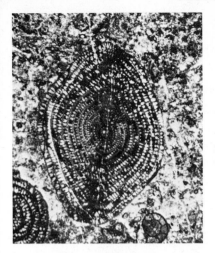

144   Internal structures in foraminifera.
Left, section of a Permian fusulinid (× 7·5). Right, section of a nummulite showing the septa
which partition the spirally coiled whorls (× 4·2); *Nummulites*, Bracklesham Beds, Eocene,
Isle of Wight.

Geological distribution

Foraminifera range from the Ordovician to the present day. Most of the
Palaeozoic forms had an agglutinated test. Calcareous forms are common
in limestones of the Upper Palaeozoic, for example in the Lower
Carboniferous as in Britain, and also in marine limestones of the Permian which occur in areas like Central Russia and parts of the United
States where fusulinids, forms like wheat grains (fig. 143j), are an
important constituent of the rocks in which they occur. They are readily
identified in thin sections of rocks, and are of great value stratigraphically
(fig. 144, left).

Planktonic Foraminifera did not appear until the Mesozoic, and they
are relatively common in the Chalk where globigerinids occur.

The nummulites were 'giant' (benthonic) forms (5–20 mm) which
characterised, and often contributed massively to, the Lower Tertiary
deposits of the Tethys region, ranging eastwards from the Mediterranean
area to the East Indies. Several species of *Nummulites* occur in the
Eocene rocks of the Hampshire Basin (fig. 144, right).

Economic use

In the search for oil many hundreds of thousands of feet of sedimentary
rock are penetrated by drilling each year. It is important that such

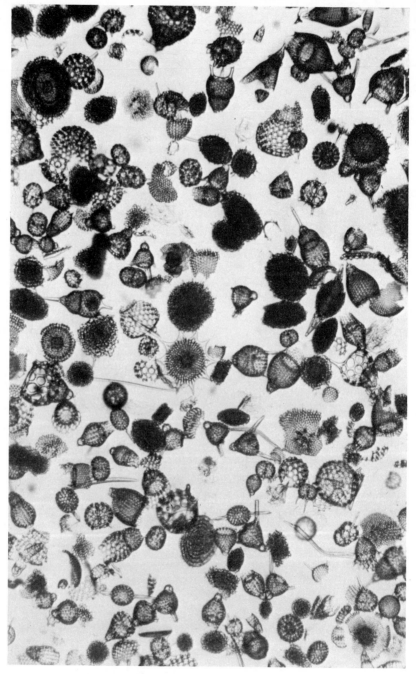

145   Radiolarians, Upper Eocene, Barbados (×70).

drilling should be stratigraphically controlled, that is to say, at any time the geologist must be able to identify the stratigraphical age of the rock being encountered in the borehole. The value of fossils for this purpose is obvious. However, the technique of drilling involves tearing the rock at the bottom of the hole into small chips which are then flushed to the surface. Any macrofossils are, therefore, macerated in this process and their value is correspondingly reduced. The small size of Foraminifera, however, means that many specimens may be obtained from one small rock chip perhaps no more than one centimetre across. It so happens that the majority of oil deposits are in the younger sedimentary rocks and in marine facies. Foraminifera are often abundant in these rocks and since many species have a short geological range they are of very considerable value in oil exploration and also in oil-field development.

Radiolaria

The Radiolaria (Cambrian to Recent) are marine planktonic organisms which secrete a symmetrical skeleton with an elaborate latticework composed typically of silica (figs. 143e, f, 145).

At the present day they have a wide distribution but their remains are conspicuous mainly in the deeper parts of the ocean where calcareous shells such as Foraminifera dissolve. Fossil forms are found in cherts and siliceous limestones of various ages, e.g. in the Ordovician of Girvan (Southern Uplands). Such cherts are not necessarily very deep-sea deposits, though a radiolarian earth of Miocene age in Barbados is possibly an authentic example.

CONODONTS

Conodonts (fig. 143l) are small tooth-like structures composed of calcium phosphate and are of unknown affinity. They range from Ordovician to Trias, and occur most usually in marine shales. They are useful stratigraphically.

17

Vertebrata, fish

The vertebrates, or backboned animals have a jointed internal skeleton of bone, or cartilage, the cardinal feature of which is the brain-case (CRANIUM), a box-like structure enclosing the brain. The group includes a great diversity of animals but they can be grouped broadly into AQUATIC forms, i.e. fish, and TERRESTRIAL forms, i.e. amphibians, reptiles, birds and mammals all of which have two pairs of 'walking' limbs and may be grouped as 'tetrapods'.

The vertebrates are assigned to the phylum Chordata along with some less familiar soft-bodied animals, e.g. sea squirts. The chordates show bilateral symmetry, and their distinctive features include a NOTOCHORD, GILL SLITS (which may be present in the adult or may occur only at an early stage in the development of the individual), and a DORSAL NERVE CORD. The notochord is an axial rod composed of soft tissue encased by a tough sheath which primitively forms a flexible support for the body. It persists as such in some lower chordates; in the vertebrates it is the basis of the backbone by which it may be replaced in development. The gill slits, paired openings in the pharynx, form part of a food filtering device in lower chordates, but in vertebrates they are either concerned with breathing as in fish, or occur only in the embryo as in tetrapods. The nerve cord is contained lengthwise within the backbone and lies above (i.e. dorsally to) the notochord.

The vertebrates comprise the largest and most varied division of the chordates, and alone have a fossil record. This begins in the Middle Ordovician in Colorado, with bits of bony plates of indeterminate fish, and continues in the late Silurian with reasonably well preserved fossils of jawless fish.

SKELETON. The skeleton may consist of CARTILAGE, a translucent organic substance which decays readily, or of BONE (p. 8) which, since it is about three-fifths mineral salt, is harder and more resistant to decay. TEETH (p. 8), which are an inconspicuous part of the body, are the most resistant part of the skeleton, being composed almost entirely of calcium salts, and may often be the only part preserved.

The skeleton is largely internal (endoskeleton) and consists of many bones which are grouped into the AXIAL SKELETON, the SKULL, and the PAIRED LIMBS and LIMB GIRDLES (fig. 146a). There is also an external skeleton of scales, or bony plates in fish, amphibians and reptiles, of feathers in birds and of hair in mammals.

The axial skeleton is made up of a series of articulating bones, the VERTEBRAE (fig. 146b, c) forming a flexible support, the BACKBONE, which is situated dorsally along the length of the body (fig. 146a). Each vertebra has a central part, the CENTRUM which primitively forms a ring round the notochord (fig. 146b and d) and in higher forms constricts (fig. 146e) or replaces it. Two processes on the dorsal side unite

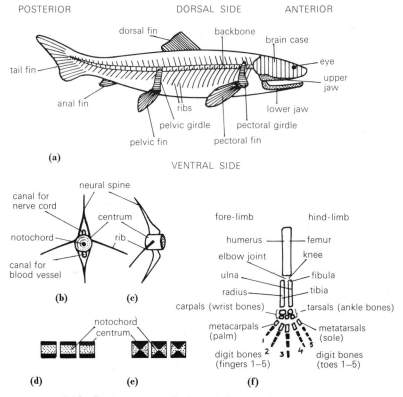

146   Basic structural plan of the vertebrate skeleton

a–e fish; a, distribution of the main parts of the skeleton; b, c, vertebrae from the tail region, seen from the front (b), and from the side (c); d, e, longitudinal sections of ring-shaped centra with unconstricted notochord (d), and biconcave centra with constricted notochord (e). f, the disposition of the limb bones in a tetrapod.

above the nerve cord forming a spiny projection (neural spine). Similar processes on the ventral side of the tail vertebrae protect blood-vessels, and articulating outgrowths from the sides form the ribs (fig. 146b, c). The SKULL articulates with the front vertebra of the backbone. It includes the brain-case and jaws.

Most vertebrates have two pairs of appendages used for steering or movement, FINS in fish (fig. 146a) and LIMBS in tetrapods (fig. 146f). The forelimbs articulate with the shoulder or pectoral girdle and the hind limbs with the hip or pelvic girdle.

FISH

'Fish' is used here in a broad sense to include four classes of aquatic vertebrates: jawless fish, the AGNATHA; primitive jawed fish, the extinct PLACODERMI; sharks, the CHONDRICHTHYES; and the bony fish, the OSTEICHTHYES. They breathe by gills, have scales covering the body and their paired appendages are fins.

Typically, the fish body is a pliant, streamlined spindle-shape, tapering tailwards and slightly flattened laterally. It is an ideal design for active swimming. The tail is the main swimming organ, and the paired fins are used in steering or, in higher fish, for braking; unpaired fins act as stabilisers. Forms which are adapted to a sedentary life on the sea-bed, e.g. the skate or sole, have a depressed flattened body with the eyes placed dorsally on top of the head instead of on each side as in active forms like the herring.

Jawless fish
Agnatha

The Agnatha include the simplest 'fish' which have no jaws, and typically lack paired fins; they have many paired gill openings. They are represented today by only a few species including the lamprey, a degenerate eel-like parasite with a skeleton of cartilage and the body naked of scales. Fossil forms, referred to broadly as the OSTRACO-DERMS, include the earliest fossils of vertebrates.

The ostracoderms were mostly heavily armoured forms with thick bony plates and scales. The endoskeleton is not usually preserved. They occur mainly in fresh- and brackish-water deposits and range in age from late Silurian to the end of the Devonian, and are most abundant in the Lower Old Red Sandstone.

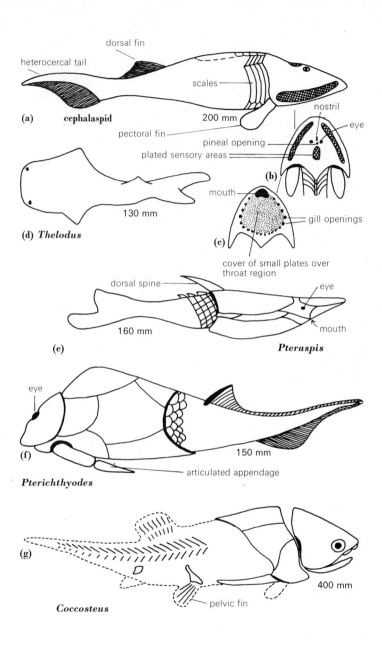

147   Old Red Sandstone fish.
a–e, ostracoderms; f, g, placoderms.

a

b

148 Cephalaspids.

a, *Cephalaspis*, Lower Old Red Sandstone, Scotland ($\times$ 0·6); dorsal view; much of the bony plating of the head shield is absent, and fine tubules can be seen radiating towards the left margin. b, *Aceraspis robustus*, Downtonian, Lower Old Red Sandstone, Røingerike, Norway ($\times$ 0·8); lateral view of a nearly complete fish with scales and fins (pectoral, dorsal and tail) preserved.

Cephalaspis (fig. 147a–c) is one of the best known ostracoderms. The head region is covered by a rigid bony shield, and the rest of the body by elongated scales. It has, unlike most of this group, a pair of fins lying just behind the head-shield. The depressed shape of the head-shield with the eyes close together on top are typical of a bottom-dwelling fish. The position of the mouth on the underside, with paired gill slits ranged behind it, is shown in fig. 147c. Dissection of the cranium in some particularly well-preserved fossils has shown that they possessed a brain structure generally similar to that of the modern lamprey. Another point of interest is the traces of tubules (fig. 148a) that run towards three areas covered with small plates which lie, one behind the eyes, and the other two near the lateral margins of the head-shield (fig. 147b). It

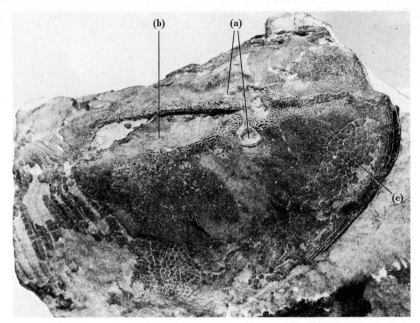

149   Cephalaspid head shield. *Aceraspis robustus*, Downtonian, Lower Old Red Sandstone, Norway (×1·5).

Position of the eyes (a), and of the dorsal sensory field (b); c, small plates covering the lateral sensory field.

has been suggested that these areas had some special sensory function; possibly they were pressure or vibration receptors.

Cephalaspis is thought to have lived in fresh-water pools or streams, feeding on organic matter filtered in the gill pouches from the sediment on the stream bed.

Upper Silurian to Middle Devonian.

Other ostracoderms are *Pteraspis* (Lower Devonian, fig. 147e) and *Thelodus* (Upper Silurian to Lower Devonian, fig. 147d). The little scales which studded the body of *Thelodus* are common in the Ludlow Bone Beds.

Fish with jaws and paired limbs

The development of jaws in fish was an important event in vertebrate evolution. Together with paired limbs and lungs, it widened the possibilities of water as an environment, and eventually led to the emergence on land of the tetrapods. The bones which form the jaws

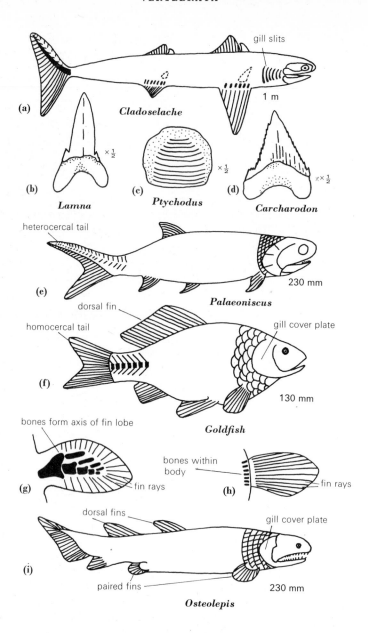

gill slits

(a)   *Cladoselache*   1 m

(b)   (c)   (d)
Lamna   *Ptychodus*   *Carcharodon*

×½   ×½   ×½

heterocercal tail

(e)   *Palaeoniscus*   230 mm

dorsal fin
homocercal tail   gill cover plate

(f)   130 mm

bones form axis of fin lobe   *Goldfish*

bones within
body   fin rays

(g)   fin rays   (h)

dorsal fins   gill cover plate

(i)

paired fins   230 mm

Osteolepis

150   Sharks and bony fish.

a–d, sharks; a, an Upper Devonian shark; b, d, teeth of predatory sharks; c, tooth adapted for crushing shells. e–i, bony fish; e, a Permian ray-finned fish with heterocercal tail; f, a modern teleost with homocercal tail; g, h, the paired fins of a crossopterygian fish, g, and of a ray-finned fish, h; i, a Middle Old Red Sandstone crossopterygian fish.

have been shown, on palaeozoological evidence, to have originated from the skeletal supports of the front gill bars.

Placoderms

The placoderms were a heterogeneous group of fossil fish possessing primitive jaws and paired limbs. They included heavily armoured forms reminiscent of the ostracoderms, and others with a less rigid scaly covering. Most were relatively small forms, but one, *Titanichthys*, reached a length of over 9 m. They included fresh-water and marine forms. They were already highly diversified when they appeared at the beginning of the Devonian and, for a time, in the Upper Devonian they were the supreme vertebrates and achieved world-wide distribution. Their numbers then fell abruptly and few survived into the Carboniferous.

Coccosteus (Middle and Upper Old Red Sandstone, fig. 147g) represents a group (the arthrodires) in which the head-shield was hinged to a second shield covering the shoulder region so that the head as well as the lower jaw may have moved to open the mouth. This fish must have had a very wide gape which, in conjunction with sharp biting bones on the jaws and a relatively streamlined body, is indicative of predatory habits.

Pterichthyodes (Middle Old Red Sandstone, fig. 147f) belongs to another group of placoderms (the antiarchs) in which a pair of spiny, jointed appendages articulated with a box-like shield of fused plates enclosing the head and trunk. Impressions of soft parts in a related genus show a pair of sacs connected with the pharynx which have been interpreted as lungs.

Sharks

The sharks and related forms, the skates and rays, form a group (Chondrichthyes) in which the skeleton consists of cartilage. Their gill slits are open, and their skin may be studded with tiny tooth-like scales, denticles. They have neither lungs nor air bladders.

Typically the sharks are predators. They are strong swimmers and their two pairs of fins are used mainly in steering. The mouth opens on the underside of the head and the well-developed jaws are typically armed with sharp pointed teeth (fig. 150b,d), but in a number of forms the teeth have a blunt surface adapted to crushing shells (fig. 150c). The skates are bottom dwellers with the body flattened and greatly extended sideways by enlarged (pectoral) fins. They feed mainly on molluscs and have teeth with a flat or grooved surface.

151   A holostean fish. *Dapedius*, Lower Lias, Jurassic ($\times$0·5).

Since cartilage is rarely preserved in fossils, the shark record consists mainly of teeth and fin spines and, apart from some early fresh-water forms, is found mainly in marine deposits. They appeared in the Middle Devonian and their numbers rose greatly during the Carboniferous and Permian. Few, however, survived into the Mesozoic and the numbers of genera subsequently has remained small. A number of surviving forms, e.g. *Lamna* (fig. 150b) and *Carcharodon* (fig. 150d) have fossil representatives in the Cretaceous and Tertiary. The number of living species is small though individuals are numerous and widespread.

Bony fish

The bony fish (Osteichthyes) are the most complex fish being distinguished by their bony skeleton and scales, and by the bony flap covering the gills. Most possess an air bladder or, more primitively, lungs. The skull is covered by a complex arrangement of plates, and the body by scales.

Bony fish far outnumber other types of fish and are enormously varied in form and habit. Most live either in the sea or in fresh water, but a few, like the eel and salmon, migrate from the one environment to the other. The two main groups into which the bony fish are separated, i.e. the ray-finned Actinopterygii and the lobe-finned Sarcopterygii were

clearly defined at the start of their history. The ray-finned fish have had
an expansionist evolution, and comprise the bulk of living fish like the
herring, perch and salmon as well as many fossil forms. By contrast the
lobe-finned fish have, since the Devonian, remained numerically
restricted and only four forms survive; their interest is primarily concerned with their role in tetrapod evolution.

RAY-FINNED FISH. In these fish (the Actinopterygii), the paired fins
each resemble a fan, consisting of a web of skin supported only by
horny rays, and with the skeletal bones lying inside the body (fig. 150h).
The fins are used for braking as well as steering, and there is a single
dorsal stabilising fin. Typically they breathe by gills. Most living
members have an AIR BLADDER, a buoyancy control which had its
origin in functional lungs. Traces of LUNGS have been found in some
fossil forms, and lungs occur in two primitive surviving members.

The earliest fossil ray-finned fish, occurring in the Middle Old Red
Sandstone, were small active forms with the head covered by bony plates,
and the body by thick, shiny, enamel-covered scales of rhombic shape
(ganoid scales). The tail axis was tilted upwards and had the fin on the
underside (HETEROCERCAL condition, fig. 150e). Ray-finned fish
remained insignificant during the Palaeozoic. In the Mesozoic a rich
variety of more advanced forms, HOLOSTEANS, appeared in both
marine and fresh-water deposits. In these, the skeleton was more completely bony and the jaws and the tail axis were shorter. The scales,
however, were still relatively thick bony plates with a shiny enamel cover
(fig. 151). By the later Cretaceous the holosteans in their turn were
largely replaced by the 'modern' type of bony fish, the TELEOSTS,
which are the dominant bony fish in both sea and fresh water today

152   A teleost fish. *Sparnodus*, Upper Eocene, Verona (×0·7).

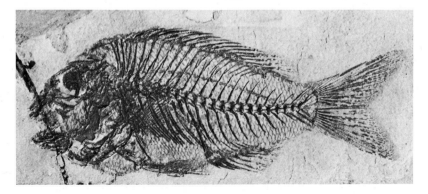

153 Lungfish toothplate. *Ceratodus*, Rhaetic Bone Bed, Trias ($\times 2$).
A thickened, ridged plate borne on the roof and the floor of the mouth, along the midline, and used for crushing shellfish.

(fig. 152). In the teleosts, the skull bones lie below the skin, the body scales are very thin bony plates lacking enamel on the surface, and the tail axis is shortened so that the fin is symmetrical and terminal (HOMO-CERCAL, figs. 150f, 152).

LOBE-FINNED FISH. These fish (the Sarcopterygii) show a variety of distinctive features, two of which are of particular significance: the paired fins are each supported by a scale-covered fleshy lobe with an axial bony skeleton (fig. 150g), and there are NOSTRILS opening in the roof of the mouth. Other characteristics, best seen in Devonian fossil members, include two dorsal fins (fig. 150i), and scales with only a thin enamel layer. Two groups are distinguished, the LUNGFISH (Dipnoi) and the CROSSOPTERYGII.

LUNGFISH. Three genera of lungfish survive today, one each in Australia, South Africa and South America. They are eel-like creatures adapted to life in water which seasonally is stagnant or liable to dry up. The Australian form gulps air at the surface of the water but cannot live out of water; the other two can survive in mud burrows until the following rainy season. Their diet, mainly of small invertebrates, is crushed by flattened teeth in the palate and floor of the mouth. Similar crushing teeth found in *Dipterus*, from the Middle Old Red Sandstone, help to distinguish this lungfish from contemporary crossopterygians. Lungfish are relatively common in the Upper Palaeozoic but are rare as fossils after the Trias (fig. 153).

CROSSOPTERYGIANS. Devonian crossopterygians differed from coeval lungfish in, for instance, having more bone in the skeleton, in the pattern of the skull bones and the nature of the teeth along the margins of the jaws. These were sharply pointed grasping teeth appropriate to a carnivorous animal. Their surface is grooved vertically as a result of radial infolding of the enamel layer into the dentine (fig. 155a). In transverse section this gives a tortuous labyrinthine pattern (fig. 155b) and the teeth are accordingly described as labyrinthodont. Such teeth are unique to crossopterygians and to certain tetrapods, the labyrinthodont amphibians, p. 254. In a number of Devonian crossopterygians the centrum of each vertebra is not ring- or spool-shaped as in other fish, but consists of two sets of bony structures. A similar construction is found in the labyrinthodont amphibians (fig. 155f) and it persists, in a modified, much more complex form, in the higher tetrapods.

The crossopterygians appeared in the Middle Devonian and were common in fresh water until the end of the Palaeozoic (fig. 154). *Osteolepis* (fig. 150i) is a typical Middle Old Red Sandstone example. Two main stocks diverged in the course of the Devonian. One led to the TETRAPODS which emerged on land and the other to the COELACANTHS which migrated to the sea. Coelacanths are common fossils in Mesozoic rocks till the end of the Cretaceous but are unknown in later rocks. Their continuing existence, however, was demonstrated by the discovery of *Latimeria* in 1938 in deep water off Madagascar.

154   A fossil shoal of crossopterygian fish from the Upper Old Red Sandstone.
Holoptychius flemingi, **Dura Den, Fife (× 0·4).**

18

Vertebrata, tetrapods

Tetrapods comprise the essentially terrestrial vertebrates characterised by lungs and typically by two pairs of limbs each with five digits. They are divided into four classes: amphibians, reptiles, birds and mammals. ORIGINS OF THE TETRAPODS. The crossopterygians (p. 251) themselves remained aquatic animals but they had, in their lungs and the structure of their bony axial skeleton and paired limbs, the necessary potential for life out of water. They are accompanied in late Devonian deposits by the earliest remains of tetrapods which resembled them in so many features that their close relationship is undoubted. The arrangement and structure of their skeletal parts is basically the same. Details of particular interest which may be closely matched include the pattern of the skull bones, the nostrils and the nature of the teeth. Structural contrasts in the skeleton of fish and amphibia reflect mechanical adjustments needed to support the extra weight of the body when it was no longer buoyed up by water. Thus, the vertebrae which in fish (p. 241) have more or less ring-like centra, in tetrapods are stout interlocking pieces which form a strong backbone. The limbs and girdles, too, were modified and strengthened to support the backbone and carry the body clear of the ground. The girdles are larger, and the pelvic girdle is united with the backbone; the limb bones are longer and fewer in number, and there are additional bones, the digits, which support the fingers and toes.

AMPHIBIA

The amphibia are the simplest and the oldest class of tetrapods. Living forms are cold-blooded and have a moist skin. They are terrestrial only in part. Typically they must go back to water for breeding, since their eggs lack protection against desiccation. Living forms, like newts, frogs and toads, are highly specialised and show little outward resemblance to their Palaeozoic ancestors.

Perhaps the most familiar amphibian is the frog. It feeds on insects and worms. Its young are completely aquatic tadpoles, with feathery gills and a long finned tail for swimming. In growth the tail and gills are resorbed, lungs and paired limbs develop and, when its metamorphosis is complete, it comes out on land as a tiny frog.

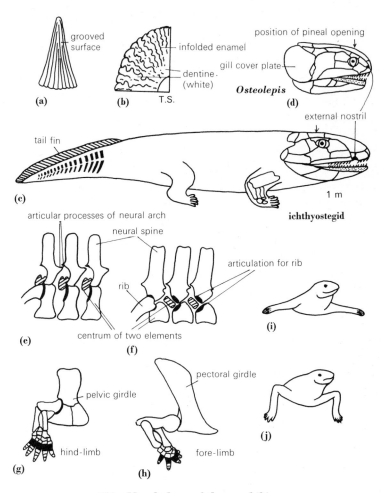

155 Morphology of the amphibians.

a, b, labyrinthodont tooth; lateral view, a, and part of a transverse section, b. c, restoration of a primitive amphibian; the skull bones are sketched in to show the similarity in their arrangement with that in the crossopterygian skull above, d. e, f, lateral view of vertebrae of an ichthyostegid (e), and of a later, more advanced amphibian (f). g, h, limb structure of a labyrinthodont amphibian. i, j, sketches to show the change in attitude of the limbs from the sprawling fish-line position in (i), to the condition in (j), typical of the early tetrapod, with the body supported clear of the ground.

156   Restoration of an ichthyostegid (× 0·1). (Painting by M. Wilson.)

The first known amphibia are the ICHTHYOSTEGIDS, which occur
in fresh-water deposits of late Devonian age in Greenland. The ichthyostegids were sprawling creatures (fig. 155c) showing a mixture of fish-
like and amphibian features. They had a fish-like tail, and vertebrae
quite similar to those in crossopterygians, but their limbs were of
tetrapod type with five digits. The arrangement of their skull bones
differed only in a few details from that in crossopterygians, their eye
openings were closer together and set higher on the skull, and they had
no gill cover plates (fig. 155c, d).

Amphibians were relatively common in the Carboniferous, more
especially in Coal Measure deposits. Many belonged to a group, the
LABYRINTHODONTS, so called from the complex infolded pattern of
the enamel layer of their teeth as seen in transverse section (fig. 155b), a
pattern also found in the crossopterygians (p. 251). They were notable
too, for the heavy, bony, armour plating of the skull (fig. 157). This
group eventually disappeared in the Permian. Amphibians are rare as
fossils in Mesozoic and Tertiary rocks. 'Frogs' and 'toads' appeared
during the Mesozoic and relatively little change in structure has occurred
in them since the Mesozoic.

REPTILES

Reptiles are cold-blooded tetrapods in which the body is protected by a dry, horny, often scaly, skin. They are emancipated from water as a breeding medium. Fertilisation is internal, and their eggs, large yolked and protected by shell, are laid on land. On hatching, the young reptile is a tiny replica of the adult.

The existing snakes and lizards, turtles and tortoises, crocodiles, and the unique New Zealand tuatara are the few relics of the highly varied array of reptiles, including dinosaurs, ichthyosaurs, pterosaurs and many others whose fossils are so distinctive of the Permian and Mesozoic rocks.

Living, and most fossil, reptiles are easily distinguished from amphibia by many skeletal details, especially in the skull, vertebrae, and limbs, and also in the soft parts. One fossil, however (*Seymouria*, Lower Permian), shows both reptilian and amphibian features so combined that its true affinity has been difficult to resolve. It was a sprawling lizard-like creature which resembled the labyrinthodont amphibia in the nature

157   A Permian labyrinthodont amphibian. *Eryops,* Red Beds, Lower Permian Texas (about $\times 0.06$).

Background; a restoration by Dr J. K. Ingham. Foreground; a replica, of a skeleton in the American Museum of Natural History, exhibited in the Hunterian Museum. Length of original, about 2 m.

of its teeth and parts of the skull, but its limbs, vertebrae, and the way in which its lower jaw articulated with the skull were more reptilian in character. While *Seymouria* indicates a close relationship between the amphibia and the early reptiles and shows some of the skeletal changes that must have occurred as the reptiles diverged from the amphibia, it is not the 'ancestral' link between the two groups since undoubted reptiles occur earlier in the Upper Carboniferous.

Early reptiles

'STEM' REPTILES. The earliest and most primitive reptiles are the cotylosaurs. They appeared in the Upper Carboniferous and dispersed widely throughout the world during the Permian. In Britain they are known chiefly from their footprints, but occasional bones and moulds have been found, for instance in Permian sandstones near Elgin, North-east Scotland (fig. 158). They were varied in type and included both carnivorous forms and plant-eaters. They are sometimes described as the 'stem' reptiles, for they are the basic stock from which the other varied types of reptiles arose.

MAMMAL-LIKE REPTILES. The advent of the 'stem' reptiles was followed rapidly, in the later Carboniferous and early Permian, by a radiation of reptiles into widely different habitats including, as in the case of the mesosaurs (fig. 159), life in water. Of particular interest amongst these early reptiles were the PELYCOSAURS. One member, *Dimetrodon* (fig. 160d) from the Lower Permian of Texas, shows an unusual development of greatly elongated spines which project from the backbone and may have supported a web of skin. The pelycosaurs are a further source of interest as the ancestral stock from which the THERAPSIDS (reptiles showing mammal-like features) and eventually the mammals themselves, differentiated.

THERAPSIDS. These mammal-like reptiles dispersed all over the world during the later Permian and Trias, their remains being especially well known from the Karroo sediments in South Africa. In the course of their history they developed, to a varying extent, skeletal features otherwise confined to mammals. Some were plant-eaters, heavy limbed and sprawling. Others, more mammal-like, were active carnivores of relatively slender build with the limbs arranged in a more or less vertical plane (fig. 160e). Unlike the typical reptile in which the teeth are all alike, the more advanced mammal-like reptiles, the THERIODONTS, had both biting and chewing teeth (fig. 160b) and they probably cut their food into small bits instead of swallowing it whole in reptile

158   A restoration of *Elginia*, a cotylosaur reptile from the Upper Permian
of Elgin, Scotland, 760 mm (about ×0·2).
(Painting by L. Kennedy and J. K. Ingham.)

fashion. They had, too, a separate nasal passage formed by the growth
of a secondary palate under the original roof of the mouth. This would
have facilitated breathing while eating. The lower jaw, as in other
reptiles contained several bones on each side, but one, the dentary was
enlarged at the expense of the other bones (fig. 160b). In mammals, the
dentary is the only element present (fig. 160c). The most advanced

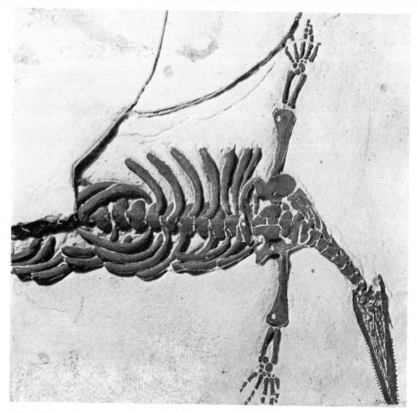

159 *Mesosaurus*, an aquatic reptile of early Permian age from the Karroo Series, South Africa. Replica ($\times 0{\cdot}7$).

The first evidence for continental drift was largely geological, and included the distribution of certain fossils such as *Mesosaurus*, which is known only from fresh-water deposits occurring on opposite sides of the South Atlantic in parts of South Africa and South Brazil.

mammal-like reptiles, the ICTIDOSAURS of late Trias age, were little creatures which are only classed as reptile by the nature of their jaw articulation. With their skeletal features so nearly mammalian, there seems a distinct possibility that these animals may also have been warm-blooded. Certainly they co-existed with the earliest of the true mammals.

The age of reptiles

From our point of view, the emergence of the mammals from the mammal-like reptiles might be regarded as the culmination of the primary radiation of the primitive cotylosaur reptiles. It was, none the less, an aberrant and relatively minor event in reptile history. The major part of their chronicle is concerned with the development of some of the most spectacular animals of all time (the dinosaurs, pterosaurs and marine reptiles of the Jurassic and Cretaceous) and also of those lesser reptiles which, insignificant in the Mesozoic, have persisted to the present day.

Dinosaurs

Dinosaur is a collective term for two separate orders of Mesozoic reptiles, the SAURISCHIA and the ORNITHISCHIA. They are distinguished by a variety of characters, including the nature of the pelvis.

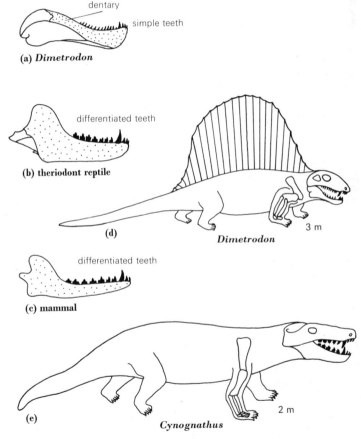

160 Mammal-like reptiles.

a–c, comparison of the lower jaws of mammal-like reptiles (a, b), with a primitive mammal's (c); note the increase in size of the dentary bone (stippled), and differences in the teeth. d, e, restorations of two mammal-like reptiles.

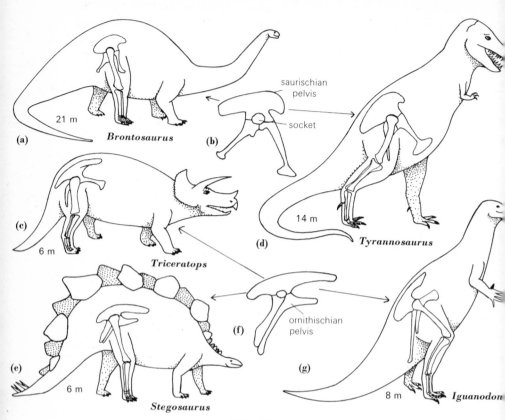

161 Dinosaurs.

The saurischian pelvis is of normal reptile type, and the ornithischian pelvis resembles that of a bird (fig. 161b,f). They included some of the largest land animals that have existed and also some that were very small.

Dinosaurs appeared in the later part of the Trias and they were the dominant tetrapods of the Jurassic and Cretaceous, and existed in large numbers until late in the Cretaceous. Their ancestry can be traced directly to a Triassic group, the THECODONTS, which also gave rise to the crocodiles, the flying reptiles and birds. The thecodonts included small, slender, carnivorous reptiles with sharp teeth set in sockets along the margin of the jaws. They were bipedal, i.e. they walked on their hind legs, and the weight of the body was pivoted on the hip bones and balanced by a long tail. The forelimbs were shorter and were adapted for grasping.

SAURISCHIANS. The saurischians include bipedal carnivorous forms, the THEROPODA, and others which reverted to all fours. (i.e. were quadripedal) and were herbivorous, the SAUROPODA. The THEROPODS (Upper Trias to Cretaceous) range from small forms, which from their build must have been capable of swift movement, to the largest terrestrial killers known to have existed. One of these, *Tyrannosaurus* (fig. 162) was a Cretaceous form which stood about 6 m high, with a massive skull over 1 m in length. The jaws were armed with dagger-like teeth suitable for tearing flesh from large animals. The forelimbs were too short to reach the mouth.

The SAUROPODS (Jurassic and Cretaceous) include not merely the largest dinosaurs, but also the largest known land animals, which it is estimated may have weighed between 30 and 50 tonnes. The record for length (26·6 m) is held by *Diplodocus* (Upper Jurassic), but its skeleton was less massively built than that of *Brontosaurus* (Upper Jurassic, fig. 161a). Typically, in the sauropods, the neck and tail were extremely long, and the body relatively short and bulky. They had stout columnar legs with broad feet like an elephant's. The skull was very small, only about 60 cm long in *Diplodocus*, and the brain capacity unusually small.

162   *Tyrannosaurus*. A reconstruction of a skeleton from the Upper Cretaceous of Montana; overall length, about 12 m.

163   *Iguanodon*. **A reconstruction of the skeleton. Height, about 4 m.**

There was however, an enlargement of the spinal cord, several times bigger than the brain, in the pelvic region. The jaws were small and, in *Diplodocus* contained a few peg-like teeth which could have dealt with only soft food. Some skeletal features, for instance the position of the eyes and nostrils high up on the skull, suggest that many sauropods may

164   *Iguanodon*. A restoration of the life appearance. Overall length, about 7 m.

have been amphibious, living for part of their time in swamps or lakes where the buoyant effect of the water would have reduced the otherwise colossal weight borne by the legs. They would have found in an aquatic environment, too, an ample supply of soft plants to crop, and security from attack by their carnivorous theropod cousins. They must, of

165   Reptile footprints from a quarry in Purbeck Limestone at Langton
Matravers, Dorset, and now on display in the Hunterian Museum, Glasgow.
The track, attributed to *Iguanodon*, was found a few feet above the Middle Purbeck Cinder
Bed and is thus of basal Cretaceous age.

course, have come ashore to lay their eggs, and massive sauropod footprints have been found in Lower Cretaceous rocks in Texas.

ORNITHISCHIANS. The ornithischians were herbivorous dinosaurs
including bipedal forms, the ORNITHOPODS, and four-footed forms,

(a) The footprints shown in fig. 165 were made by a large bipedal dinosaur, probably *Iguanodon*. Diagram (a) shows how closely the right foot skeleton fits into one of the footprints.

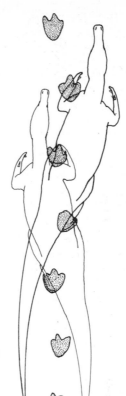

(b) The disposition of the footprints sheds some light on the way *Iguanodon* walked. The impressions, though clearly alternate, are almost in line. This suggests that at each stride the body swayed from side to side, the head and tail region remaining parallel to the direction of movement. This kind of movement involves some rotation of the foot relative to the body which would normally take place at the ankle. The absence of a tail groove indicates that the tail was held clear of the ground to counterbalance the body.

166   Reptile gait; an explanation of how *Iguanodon* walked.

Diagram redrawn from a photograph. The explanation, by Dr J. K. Ingham, accompanies the exhibit of the foot tracks shown in the previous figure.

such as the STEGOSAURS and the CERATOPSIANS. By dinosaur standards, none were of gigantic size.

The ORNITHOPODS appeared in the Upper Jurassic, *Iguanodon* (Lower Cretaceous, figs. 163, 164) being one of the best known. It was a large kangaroo-like creature, standing about 4·5 m high. It had a horny beak and large grinding teeth along the sides of its jaws. Its feet, with the number of toes reduced to three, were similar to those of a running bird like the ostrich (see fig. 166, reptile gait). *Iguanodon* bones and footprints (fig. 165) occur in Wealden beds in South-east England, for instance, in Sussex and the Isle of Wight. Its fossilised teeth were

167 *Stegosaurus*; a restoration of the life appearance. Length of original
about 9 m.

the first dinosaur remains to be discovered and recognised as reptilian in
origin; they were found and identified as reptile teeth by Gideon
Mantell in 1822 in Sussex.

Four-footed ornithischians appeared in the Lower Jurassic but were
more characteristic of the Cretaceous. There are several groups, varied
in appearance but all retaining a trace of their bipedal ancestry in their
shorter forelimbs. Some Cretaceous forms were heavily armoured with
bony plates covering their backs much as in the modern mammal, the
armadillo. Others like the stegosaurs and the ceratopsians were less
heavily armoured.

The STEGOSAURS were mainly Jurassic forms. *Stegosaurus* (fig. 167),
for instance, had two rows of erect bony plates along the backbone and
paired spikes on its stumpy tail. The CERATOPSIANS did not appear
until the Upper Cretaceous. *Triceratops* (fig. 161c) had a bony frill
projecting over its neck from the back of the skull and bony horns, like

168   A ceratopsid; a restoration of the life appearance of an adult *Procera-tops* along with newly hatched young forms. Length of original about 2 m.

169   Dinosaur eggs; a nest of eggs found along with the remains of a
        ceratopsid dinosaur (*Proceratops*) in Mongolia.

170 *Cryptocleidus:* a reconstruction of a long-necked plesiosaur skeleton

those of a rhinoceros, on its nose and forehead. One ceratopsian is known in unusual detail, having been found in varying stages of development, from fragmentary embryos inside eggs to the fully developed adult stage, in Cretaceous rocks in Mongolia (fig. 168).

Marine reptiles

Several different groups of reptiles were adapted in varying degrees to life in water. They showed changes in body shape and limb structure which enhanced their capacity for swimming but left them recognisably reptiles. Two groups, the PLESIOSAURS and the ICHTHYOSAURS, which are relatively common as fossils, illustrate some of these modifications in form.

PLESIOSAURS (figs. 170, 171) were relatively large reptiles ranging from about 3 m to over 12 m in length in later members. A typical form is *Plesiosaurus* (Jurassic) which had a short, almost barrel-like body with a very long neck and small head. The limbs (fig. 172b) were large flippers with the upper limb bones elongated, and the digit bones greatly increased in number, in some cases to over 12 joints in each digit. It is probable that plesiosaurs lived for most of the time in the sea, and swam using their flippers as hydrofoils as do turtles; almost certainly,

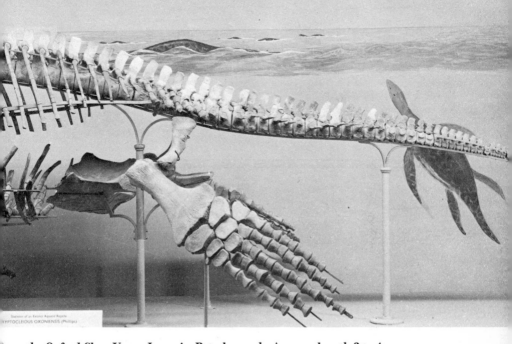

Skeleton of an Extinct Aquatic Reptile
KYPTOCLEIDUS OXONIENSIS (Phillips)

from the Oxford Clay, Upper Jurassic, Peterborough. Average length 3 to 4 m.

however, they returned to the shore for breeding. Plesiosaurs were widespread in the Jurassic and Cretaceous, and are found throughout rocks of this age in Britain.

ICHTHYOSAURS were perhaps the most specialised of all the aquatic reptiles and, judging from their fish-like build, were able to swim at high speed. *Ichthyosaurus* (Jurassic, fig. 172a) a typical example, was about 3 m in length. Well-preserved specimens in dark shale may show the outline of the body (fig. 173). It was spindle-shaped, like that of a dolphin, with a dorsal fin, and a tail fin above the down-turned axis of the tail. The limbs were reduced to short steering and braking 'fins', each with very many small disc-like bones (figs. 172c, 174). The skull had long pointed jaws with sharp conical teeth, the eyes were large, and the nostrils placed high on the head (fig. 172a). Ichthyosaurs are thought, like modern whales, to have given birth to live young at sea since occasional fossils have been found showing very small ichthyosaur skeletons within the body cavity.

Ichthyosaurs range through most of the Mesozoic. Their remains are common in some horizons in Britain, between the Jurassic and Middle Cretaceous. Occasionally more or less complete skeletons occur but isolated bones, such as vertebrae, or teeth are more frequent.

171   *Cryptocleidus*; a restoration. Length, 4 m. (Model by Mr F. Munro
and Dr J.K. Ingham.)

Flying reptiles and birds

The earliest tetrapods capable of flight belong to two groups, the flying
reptiles, or PTEROSAURS, and the BIRDS. They are believed to have
developed independently from the Triassic thecodont stock from which
the dinosaurs also diverged (p. 260), and they have a number of features
in common. For example, in both groups, the forelimbs were modified
to form wings, and the breast-bone was developed to form a surface for
the attachment of the flight muscles. The pterosaurs had delicate bones,
some of which were hollow and thin walled, thus achieving a light
skeleton without sacrificing strength. Birds, too, except for the earliest
one, have hollow bones (filled, in living forms, with air sacs). Some
pterosaurs and the earliest birds had long pointed jaws with sharp teeth.

However, despite these and certain other reptilian features, the birds
are a very uniform group, sufficiently distinct to merit their being
assigned to a separate class, Aves. In particular, they differ in wing
structure (fig. 175), in having an insulating cover of feathers, and in
maintaining a constant warm body temperature. There is no evidence
concerning the body temperature in pterosaurs though it may be con-
jectured that the great expenditure of energy involved in flight may have
required a warm-blooded condition.

PTEROSAURS. Some of the typical pterosaur features are illustrated by

Rhamphorynchus (Upper Jurassic, fig. 175a). It had a relatively small body with a long tail at the end of which was a rudder-like membrane. The wings were formed by a membrane of skin stretched between the body and the forelimbs. Each wing was supported along its forward edge by the bones of the forelimb and fourth digit; the latter, especially, were greatly elongated. The first three digits, which were short and bore claws, remained free. The hind legs were long and slim, and so articulated that normal walking may have been difficult.

The earliest known fossil of a pterosaur *Dimorphodon* (fig. 176) comes

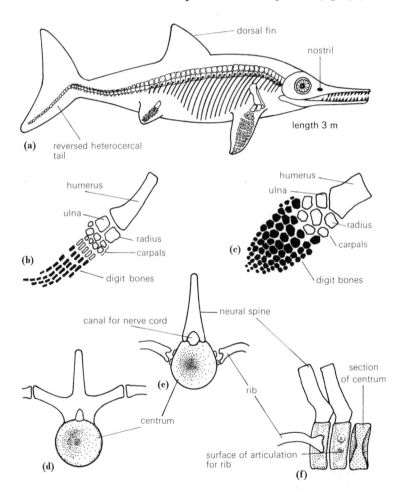

172 Marine reptiles.

a, restoration of an ichthyosaur. b, c, right fore-limbs of (b), a plesiosaur and (c), an ichthyosaur. d–f. dorsal vertebrae of (d), a plesiosaur and (e, f) an ichthyosaur.

from the Lower Lias of Lyme Regis, and the group persisted through the Jurassic and much of the Cretaceous. The pterosaurs were a varied group, ranging from sparrow-size to the gigantic *Pteranodon* with a wing-span of about 7 m. In some of the later forms, the PTERODACTYLS, the tail was much reduced and the jaws lacked teeth.

BIRDS. The birds are a varied group. Some, like the swift and albatross are almost incessantly in the air, while others like the ostrich and the penguin have lost the power of flight. They are widespread today, but their fossil record, especially of the early forms, is scanty.

The earliest known fossil bird is *Archaeopteryx* (fig. 175b) of which three finely preserved skeletons have been discovered in the Solenhofen Limestone (Upper Jurassic). They are undoubtedly skeletons of birds for they show the imprint of feathers splayed out around the bones, but there are a number of features in the skeleton which show convincingly that these creatures were derived from reptile stock.

Archaeopteryx resembles its reptilian forbears in having teeth in its jaws, three clawed fingers on each wing, no hollow bones and a long tail. The tail is a unique structure consisting of a long chain of reptile-like vertebrae with a row of feathers down each side. Bird-like characters, apart from the feathers, include a 'wish' bone, the arrangement of the elongated arm bones (which form the forward edge of the wings) and the long flight feathers extending back from these, and also the strong legs with four clawed toes arranged as in perching birds.

The next known bird fossils occur in the Upper Cretaceous, particularly in the Chalk deposits, though the only known British example consists of isolated bones found in the Cambridge Greensand. These fossils are mostly different types of sea-birds, including a tern-like bird

173 An ichthyosaur with the outline of the soft body preserved. *Stenopterygius*, Upper Lias, Lower Jurassic, Holzmaden ($\times$ 0·1).

174   Head and fore-limbs of an ichthyosaur ($\times$0·15).

capable of strong flight, and a diving bird, *Hesperornis*, which was not capable of flight. They resembled modern birds, in having a number of hollow (pneumatic) bones, in the reduction of the bony tail and, except for *Hesperornis*, in lacking teeth.

Many of the kinds of birds now living were represented in the Eocene, and the group continued to diversify during the Tertiary. Fossils are generally fragmentary but Pleistocene tar-pits and peat deposits have yielded a variety of more complete specimens.

Surviving types of reptiles

Sphenodon. *Sphenodon*, the tuatara, is the only surviving member of the rhynchocephalians, a group which was widespread in Trias times but of which later fossils are rare. *Sphenodon* is a lizard-like creature living (and protected) on islands off New Zealand. It shows little skeletal change from its Triassic forebears. A primitive feature of some interest is its pineal eye, an eye-like structure with lens and retina which lies below the skin on top of its head. The pineal eye explains the nature of an opening in the skull in a number of fossil forms, e.g. the crossopterygians, some amphibians and reptiles.

LIZARDS AND SNAKES. The commonest and most varied of modern reptiles are the lizards and snakes. They are derived from Triassic forms. Lizards are rare as fossils during the Mesozoic apart from an aquatic group, the mosasaurs, which were widespread for a time in the Cretaceous seas. Snakes are basically 'legless' lizards which appeared during the Cretaceous.

TURTLES AND TORTOISES. The turtles and tortoises include both marine and land-living forms. Their main characteristic is the box-like shell within which the body is encased and into which the head and limbs can be retracted. This shell consists of bony plates covered with horn. The group originated from cotylosaur stock during the Trias and its subsequent history is fairly well documented, its essential structures having changed but little. Fossils are relatively common in some rocks

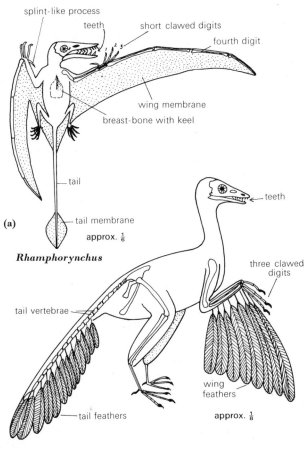

175　A flying reptile and the first bird.

176 A pterosaur in flight. *Dimorphodon*, Lias, Lower Jurassic, England; length about 1 m. (Painting by M. Wilson.)

in England, occurring for instance in the Purbeck and Wealden Beds, in the Chalk and in the London Clay.

CROCODILES. The crocodiles are amphibious, predatory reptiles which today are restricted to warm countries. Of the surviving reptiles they are the most nearly related to the dinosaurs, and they retain a trace of their origin from bipedal stock in the larger size of their hind limbs relative to the forelimbs. They appeared in the late Trias and were widely distributed in the Jurassic and Cretaceous. Forms similar to the modern crocodile appeared in the Tertiary. In England crocodile remains are found in Wealden and Purbeck Beds, and in the Eocene (fig. 178).

177 *Archaeopteryx* in 'flight'. (Model by Mr F. Munro.)

MAMMALS

Two of the diagnostic features of mammals are unique: the presence of fur or hair, and the secretion of milk as a food for their young. As in the birds, the body temperature is kept at a uniform warm level. Living varieties are adapted to a wide range of habitats on land, in water and in the air. They are assigned to three groups: the majority, including man and the hedgehog, to the placentals; a relatively small number, including the kangaroo, to the marsupials; and three genera, one of which is the platypus, to the monotremes. Fossil mammals include rare Mesozoic forms and an extensive array of Cainozoic forms which document the changes leading up to the modern mammals.

Convincing evidence of the reptilian origin of the mammals is provided by the similarities in skeletal details between the mammal-like reptiles and the mammals which have already been discussed on p. 256. The mammals differ greatly from the main body of reptiles, however, not only in the organisation of the soft tissues but also in a number of skeletal details. Foremost amongst these is the larger size of the mammal's brain-case (fig. 179h) and its indications of a more intricate brain.

178 Crocodile scute: one of a series of protective bony plates from the back of a crocodile; *Diplocynodon*, Oligocene, Hampshire (×1·5).

The nature of the lower jaws and of the teeth give possibly the simplest diagnosis and, fortunately, these are the parts most often preserved. Each half of the lower jaw consists entirely of one bone, the DENTARY, in mammals (fig. 160c) but contains several bones in reptiles (fig. 160a). A mammal has dissimilar biting and chewing teeth, the latter having divided roots, and in its lifetime grows two sets only; a juvenile, or MILK, SET, and an adult, or PERMANENT, SET. A reptile on the other hand has uniform simple teeth, each with a single root, which may be replaced indefinitely. Other points of difference include the form of the articulating surface between the skull and the backbone which in mammals is a double-headed knob, but in the typical reptile is a single knob.

Mesozoic mammals

From their first appearance in the late Trias until almost the end of the Cretaceous, mammal remains occur only sporadically and are fragmentary, consisting of jaws, teeth (fig. 180e–h) or parts of skulls but not complete skeletons. These early mammals were small, the largest being about the size of a small domestic cat. They have been referred to several orders each distinguished by the CUSP pattern of the cheek teeth. One order, the PANTOTHERES (Jurassic, fig. 180e,f) is thought to be the likely group from which both the marsupials and the placentals were derived.

The main horizons from which fossils of these mammals have been found in England include the Rhaetic (from fissures in Carboniferous Limestone in the Mendips area and Glamorgan), the Stonesfield Slate, the Purbeck 'dirt' beds and the Wealden Series.

Cainozoic mammals

MONOTREMES. The duck-billed platypus and the spiny ant-eaters are the only known monotremes; they are confined to the Australian region. They are covered with fur and, while they lay eggs, they also secrete milk. They are known as fossils only from the Pleistocene. However, they show a number of reptilian features in their skeleton which suggest they are probably a more ancient group and may have descended separately from the therapsid reptiles.

MARSUPIALS. The marsupial mother carries her young in her pouch, a fold of skin over her belly, which is supported by special bones. The young are born while still very immature, and crawl into the pouch where, attached to teats, they are fed on milk. Marsupials form a minor group today and, apart from an occasional form like the American opossum, they are confined to Australia.

The earliest fossils of marsupials are of opossum-like creatures from the Upper Cretaceous in America. A wide variety of Tertiary marsupials are recorded in South America which was for a time, in the Tertiary, cut off from North America. Little is known of their pre-Pleistocene history in Australia. Modern forms show adaptations to a wide range of habitats which are similar to those found in the placental mammals.

PLACENTALS. In placental mammals the developing embryo is nourished in its mother's uterus by a special growth, the placenta, and when born it is relatively mature; in some cases, like the whale, it may be able to follow its mother within hours of birth. The skeleton differs from that of marsupials in a variety of features; the brain-case for instance is larger, and the palate is a solid bony plate. Placentals include the vast majority of living mammals, and most (about 95%) Cainozoic forms.

Placental mammals appeared in the Upper Cretaceous about the same time as the marsupials. The earliest fossils are tiny skulls and jaws of a shrew-like insectivore. Insectivores are also the most primitive placentals, and it is probable that the Cretaceous members are the ancestral stock from which placentals differentiated. By the late Eocene most of the orders into which they are separated had become distinct. Some of these orders are now extinct, while others survive with varying degrees of strength.

The adaptive radiation of the placentals was accompanied by modifications of the skeleton which are related to three aspects in particular: development of the brain, locomotion and diet.

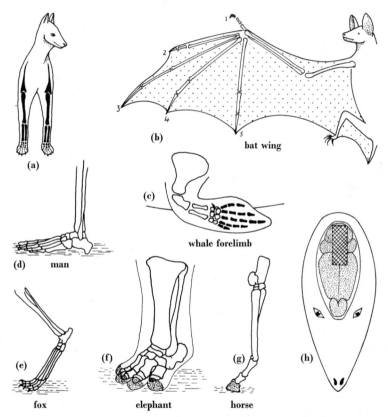

(b) bat wing

(a)

(c)

whale forelimb

(d) man

(e) fox

(f) elephant

(g) horse

(h)

179 Limb form and brain size in placental mammals.

a, diagram showing the limbs of a typical four-legged mammal, aligned vertically below the body. b, c, adaptations of the fore-limb for flight (b), and for swimming (c). d–g, different types of foot structure in walking and running mammals. h, comparison of brain size of a primitive placental mammal (stippled) with that of a reptile (cross-hatched).

The brain in primitive placentals is small and relatively simple. In more advanced forms the brain-case is, relatively, larger and it also shows an increase in complexity of the brain. The latter implies improved co-ordination of bodily functions and development of intelligence and sense organs like the eyes or ears.

The primitive mammal has short legs and walks with its five-toed feet flat on the ground as in fig. 179d. In more advanced forms the limbs and the way in which the body is carried are modified so that they became more agile. Usually there has been an elongation of some of the limb bones which has resulted in an increased length of stride, and there may also be some alteration of the arrangement of the foot bones. Thus an agile form like the fox may walk on the digits with the sole of the foot raised off the ground (fig. 179e) or a fast runner like the horse may walk on the tip of its digits (fig. 179g). The most complete structural change of the basic limb plan is seen in a flying form like the bat (fig. 179b), or an aquatic form like the whale (fig. 179c).

The nature of the teeth is of particular importance in the study of placentals and is a good guide to the relationship between different forms. It has to be remembered, however, that animals which are unrelated to one another but which eat the same food, may develop a similar tooth structure.

The maximum number of teeth in placental mammals, 44, is found in the more primitive forms which have three biting teeth, the INCISORS, one piercing tooth, the CANINE, and seven chewing or CHEEK TEETH on each side of both upper and lower jaws (fig. 180d). The incisors are simple round pegs with single roots inserted along the front margins of the jaws. The canine which is longer and pointed, is interposed between the incisors and the cheek teeth. The latter, lying towards the back of the jaw, have two or more roots, and the visible part, the crown, bears sharp projections, the CUSPS, on the grinding surface. The first four cheek teeth, the PREMOLARS, are of simpler structure than the three rear teeth, the MOLARS which typically are the most informative about diet (fig. 180i–o). In primitive forms the molars are triangular in plan with a cusp at each corner, but in the lower jaw each one has an additional small platform jutting out on the rear side (fig. 180b, c).

The dentition described above is found primarily in insectivorous (insect-eating) animals and it is modified in more advanced forms which feed on other types of food. The modifications include both a reduction in the number of teeth (e.g. to 32 in man, or as in an extreme case, the whalebone whale, to none) and an alteration in their structure. The most profound changes are seen in the cheek teeth. Thus, in carnivorous forms, like the cat, a pair of opposed teeth (i.e. one in the lower jaw and the other immediately above it in the upper jaw) have the cusps compressed and united to form a jagged cutting edge which can slice through the sinews and crush the bones of their prey. These teeth are known as CARNASSIALS (fig. 180k–m). The cat has one pair on either side of the jaws. Herbivorous forms, like the horse, have enlarged 'prismatic' cheek teeth with a high crown and complex grinding surface in which bony cement (which in other types of teeth is usually restricted to the

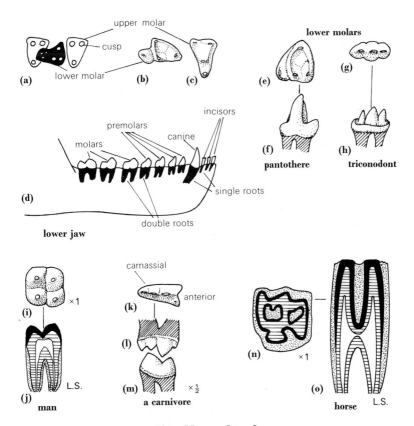

180 Mammal teeth.

a–c, teeth and d, lower jaw of a primitive (insectivorous) placental mammal (a shows how the cusps of opposed upper and lower molars alternate in position in the process of chewing; b, c, crown view of molars). e–h, molars (lower jaw) from two different types of Jurassic mammals (e, g, crown view; f, h, side view). i–o, adaptation of a molar for a general diet, (i, j); for a carnivorous diet (k–m); and for a herbivorous diet (n, o) (i, k, n, crown views; l, m, side view of an upper and a lower molar; j, o, sections of unworn molars to show the arrangement of enamel (black), dentine (lined), cement (stippled); pulp cavity (unshaded)).

roots) is incorporated (fig. 180n–o). This structure is highly resistant to the abrasive action involved in grass chewing. It may be noted that some grasses may contain a small amount of silica.

Some examples to illustrate the adaptive radiation of the placental mammals

The placentals are separated into a large number of orders. Some of these are considered briefly to show the range of structural variation involved.

INSECTIVORES. The insectivores are represented today by small, often nocturnal forms like the hedgehog, shrew and mole which feed on insects, worms or slugs. Although some forms are highly specialised for particular modes of life, they retain many of the primitive features of the skeleton shown by the early insectivores. They have, for instance, a small, relatively simple brain and primitive dentition. They have short legs and they walk with their clawed feet flat on the ground. They have a fairly continuous fossil record from their first appearance in the later Cretaceous, through the Tertiary to the present day.

BATS. The bats, an offshoot of the insectivores, are highly specialised for flying. Their fossils are understandably rare, but examples from the Eocene show that their power of flight was, even then, well developed. The form of the wing (fig. 179b) makes an interesting comparison with that of the pterosaurs and birds (fig. 175). The wing membrane is a fold of skin extending from the body as far back as the tail and supported by the long forearm and enormously elongated finger bones (except the thumb) which are arranged rather like spokes in an umbrella.

PRIMATES. The primates are typically arboreal (tree-living) animals. They are in some respects relatively unspecialised. They move on all fours, apart from a small number of ground-living forms which are bipedal. The hands and feet are adapted for grasping and, for this purpose, the thumbs and large toes are opposable (fig. 181f). All five digits are retained and their tips are protected by flat nails. In the more advanced forms the teeth are reduced in number and slightly modified for eating a variety of food. The most significant modifications in the development of the group, however, are the expansion of the brain, with consequent enlargement of the brain-case, and the improvements in eyesight.

Living primates are widespread in warm climates. They include relatively simple forms, the LEMURS (fig. 182i, j) and TARSIERS (fig. 182g, h), and more advanced forms, the ANTHROPOIDS, i.e. monkeys, apes and man (fig. 182a–f). Fossil primates are on the whole rare. The earliest forms, from the base of the Tertiary, were small arboreal animals, some of which are difficult to distinguish from insectivores.

The LEMURS are small nocturnal creatures with a long tail, and a pointed fox-like head with the eyes directed to the sides (fig. 182j). Their teeth are of primitive type with sharp cusps. Fossil lemurs occur in Eocene deposits (fig. 182i), and subsequent forms show relatively little change in structure.

The higher primates, or ANTHROPOIDS, show a number of specialised features. For instance the skull is modified by an increase in size of the brain case, and by the direction forwards of the orbits (which house the eyes). The face is shortened, too, by the reduction of the snout (due to diminished use of the sense of smell) and jaws. The smaller size of the jaws is linked with a reduction in the number of teeth to 32 in most members. The teeth are somewhat modified; the molars, for instance, are quadrate and have four or five blunt cusps, which are ideal for a generalised diet.

In the course of time monkeys, apes and man have diverged to follow quite different modes of life. Monkeys are the more remote from man.

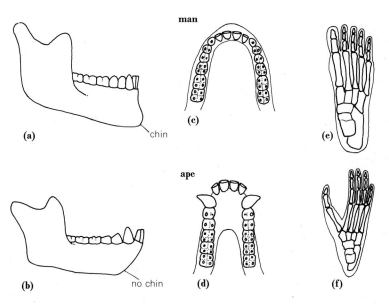

181 Points of contrast between apes and man.

a, b, profile of lower jaw. c, d, alignment of the teeth in the lower jaw. e, f, form of the foot (e, no grasping power and the toes are shortened, f, large toe opposable and the foot can grasp).

Typically they are arboreal and use both hands and feet (sometimes the tail also) as they run or leap among the trees. Apes, too (except the gorilla), live mainly in trees, where they move, swinging by their exceptionally long arms, from branch to branch. Their legs are short and they have no tail. Man in contrast, is entirely ground-living and bipedal.

While man and apes, like the chimpanzee, show marked similarities in their general anatomy and physiology, they differ significantly in certain skeletal details, particularly those connected with the development of the brain, with posture and mode of movement, and to a lesser extent with diet.

The brain capacity in man is about 1300 cm^3, compared with about 500 cm^3 in a gorilla. Some parts of the brain, too, are more complex, especially in the frontal region and this is reflected in the higher forehead in man (fig. 182b). Man has a chin (fig. 181a) and his jaws are less massive with the teeth lying in a gently rounded arch (fig. 181c) whereas apes have no chin (fig. 181b) and the teeth lie parallel on either side of the jaw (fig. 181d). Man is distinguished, too, by his erect posture with the skull balanced on top of the neck. The backbone has an S-shaped curvature in man but is more C-shaped in apes. In walking, man places the sole of the foot on the ground, and the large toe is not opposable (fig. 181e), while apes continue, to some extent, to move on all fours and their feet retain a grasping function (fig. 181f).

The earliest fossil anthropoids occur in Lower Oligocene rocks in Egypt. They are of rather generalised forms which, however, indicate that the divergence of monkeys and apes had already begun; in later rocks occasional fossils of more advanced monkeys and apes occur. The remains of a small ape '*Proconsul*', from the early Miocene in East Africa (fig. 182f), with relatively short arms and a number of other unspecialised features, suggest that at that time apes were not yet fully adapted to life in the trees, and may to some extent have been ground-living. Such forms may be close to the lineage which, by becoming fully adapted to life on the ground, led to man. There is, however, no fossil evidence of man until the Pleistocene.

The earliest fossil showing human features is *Australopithecus* (fig. 182d) from the early Pleistocene of South Africa. This fossil is of a small 120 cm 'man-ape' and gives some idea of the changes which separated man's forebears from the early apes; it represents a pre-human stage of the hominids. *Australopithecus* walked erect and had large teeth of essentially human type, but its face was more ape-like and its brain capacity between 600 and 700 cm^3.

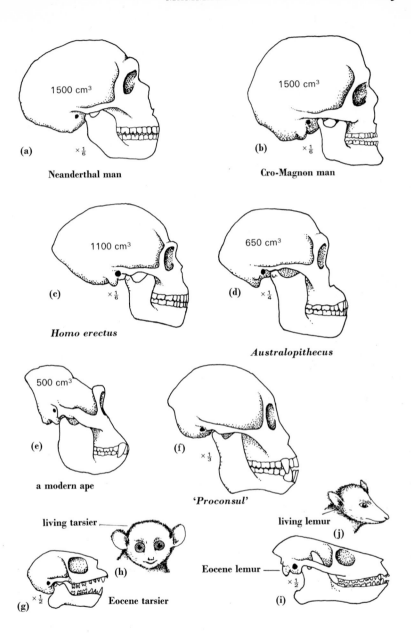

182 Primate skulls.

Human fossils are of increasingly frequent occurrence during the Pleistocene. As well as his bones, fossil evidence includes tools (artefacts), camp sites with fire ashes, broken and charred animal bones, drawings and carvings. The sequence of human cultural development is best known in Europe where artefacts have provided the basis for the establishment of successive cultural stages. The scattered nature of the evidence has made correlation a difficult matter especially over wider areas beyond Europe.

Undoubted human fossils occur widely in the Middle Pleistocene, e.g. in Java, near Peking and in Algeria. They are assigned to *Homo erectus*, formerly *Pithecanthropus* (fig. 182c). This form has a brain capacity of about 900 to 1200 cm^3, erect posture, and large but human-type teeth. Some ape-like features are seen in the skull, for instance in the pronounced brow ridges of the receding forehead and the somewhat protruding jaws without a chin. Primitive implements and fire ashes are associated with these remains.

Later fossils, sometimes entire skeletons, are characterised by having a brain capacity of about the same size, or even larger than that of modern man, *Homo sapiens*, viz. about 1450 cm^3 on average in European males. An early race of man which is particularly well known is Neanderthal man (fig. 182a). He appeared in Europe and adjacent parts of Africa and Asia in the third interglacial (Upper Pleistocene). He was of short stature (163 cm in height) and powerful build. Distinctive features which set him apart from *Homo sapiens s.s.* include the large thick-walled skull with a strongly receding forehead and massive brow ridges above the orbits. Neanderthal man was a hunter, who used fire, and made well-formed stone tools.

Fragmentary fossils suggest that *Homo sapiens* existed in the second interglacial, but his appearance en masse was during the last glaciation in Europe, about 40,000 years ago, when people of distinctly modern aspect, the CRO-MAGNONS, entered the unglaciated regions and displaced the Neanderthalers. The Cro-Magnons (fig. 182b) were tall people (men about 183 cm) with high brain capacity (average 1500 cm^3), and a high forehead, whose cultural attainments eclipsed those of their predecessors.

WHALES. A number of mammals are adapted to an aquatic life. The most specialised of these are the whales which range from the blue whale of about 30 m in length to the quite small porpoises and dolphins. They have no known terrestrial relatives. While they retain mammalian characteristics, like, for instance, their mode of reproduction and of

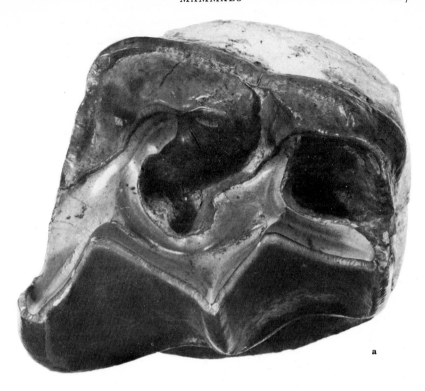

a

b

183   A mammal tooth and jaw.

a, grinding surface of a herbivore tooth; *Palaeotherium*, a perissodactyl, Bembridge Limestone, Oligocene, Hants. ($\times 2 \cdot 5$). b, lateral view of the lower jaw of a primitive carnivore; *Hyaenodon*, a creodont, Lower Headon Beds, Eocene, Hants. ($\times 0 \cdot 7$).

temperature regulation, they have developed many fish-like features which parallel those shown by the Mesozoic ichthyosaurs (p. 269). The dolphin, perhaps, comes closest to these reptiles in its appearance; it is a streamlined spindle shape with a long head which joins the body without a neck region. It swims by an up-and-down movement of the tail, which unlike the ichthyosaur's has horizontal flukes (fins). The forelimbs are short fin-like stabilisers (fig. 179c), and the hind limbs are vestigial. Its teeth are sharp spikes, ideal for grasping slippery fish.

The first whales, which occur in the Eocene, had teeth of primitive type, the front teeth being pointed and the cheek teeth having sharp cusps. By Miocene times most of the modern types of whales had appeared.

CARNIVORES. The carnivores are flesh-eating animals including cats, dogs, weasels, bears, aquatic forms like the seals, and also a great variety of extinct animals. The skeleton is relatively primitive. Typically, carnivores can move very fast for short distances in pursuit of the prey on which they feed, and their claws are adapted for attacking and holding their prey. Their most specialised feature is their dentition. For instance, in most the canines are long and sharp for stabbing prey, the cheek teeth have sharp cusps, and typically two opposed carnassials (p. 281, fig. 180k–m) are present on each side of the mouth.

Among early carnivores were the CREODONTS which appeared at the beginning of the Tertiary. They were at first little different from the insectivores but they diversified during the Eocene. Typically they had a small brain, and their limbs were short so that they were not capable of great speed. Their teeth showed some modifications to cope with a carnivorous diet, and included carnassials in later forms such as *Hyaenodon* (fig. 183b) from the late Eocene and Oligocene. They were mostly of small size, ranging from weasel to lion size. Their numbers dwindled in the Oligocene but a few survived till the Pliocene.

The creodonts were displaced towards the end of the Eocene by the miacids, a family ancestral to modern carnivores, which were faster moving and more intelligent than the archaic creodonts. The carnivores reached their acme in the Pliocene and they remain a numerous and diverse group.

UNGULATES. 'Ungulate' is a term which covers several groups of plant-eating mammals most of which have hooves. Amongst living forms are the widely different ELEPHANTS (proboscideans), HORSES (perissodactyls) and CATTLE (artiodactyls). There are also many fossil ungulates.

184   The lower jaw of a Pleistocene elephant.

On each side of this jaw is a worn-out 'grinder' (molar) which is almost displaced from the
front of the jaw by the eruption of the second, relatively unworn, grinder seen at the rear.
Elephas antiquus, Pleistocene Gravels, Barrington ($\times 0.2$).

Adaptations shown by the ungulates are largely concerned with diet
and locomotion. The teeth are generally modified, the incisors being
flattened to facilitate cropping and the cheek teeth enlarged with a high
crown and complexly ridged grinding surface to crush the plant material
(fig. 183a); the cheek teeth may continue to grow as they are worn down.
In the main, limb changes are connected with the need to escape
quickly from enemies and also with ranging over wide areas in search of
fresh pastures. Thus, for example, the limb bones are elongated and this
lengthens the stride. Also they walk on the tips of the toes only and these
are protected by horny hooves (modified claws) against jarring contact
with hard ground (fig. 179g). The side toes may be reduced or lost
except in forms like the elephant, whose great weight requires the
support of a broad foot (fig. 179f).

ARCHAIC UNGULATES. The ancestry of most, if not all ungulates is
traced from the CONDYLARTHS, forms originating from insectivores,
which appeared at the beginning of the Tertiary. The earlier members
had short limbs with clawed feet, a long tail and teeth of primitive type.
In later members the limbs were longer, hooves replaced the claws and
the cheek teeth were broader and had more cusps. These forms, and
a variety of archaic derivatives, some reaching the size of a large

rhinoceros, were on the decline in the latter part of the Eocene. Their place was taken by the ancestors of modern ungulates, forms which, being relatively more intelligent and faster moving, were better equipped to evade capture by carnivores.

ELEPHANTS. The existing elephants (*Elephas* and *Loxodonta*) show a number of unusual and highly specialised features, for instance in the structure of their teeth. These are contained in short jaws, and at any one time there is room for only four molars, one in each half-jaw. As these are worn down they are pushed forwards and out by the next set of four (fig. 184), and this happens twice during the life of the animal so that the eventual total of molars is 12. The molars are enormous, high-crowned grinders with many transverse ridges (fig. 185c, d). There is only one pair of elongated incisors, known as tusks (fig. 185a), and no canines.

The modern elephants mark the culmination of a series of changes which can be traced in an extensive fossil record starting with *Moeritherium* (fig. 185b) from the Upper Eocene in Egypt. This form, if not directly ancestral, gives a picture of the probable basic elephant stock. It was a small pig-like animal with stout legs. It had a nearly complete set of teeth (i.e. it lacked only one pair of cheek teeth in each jaw, and one pair of incisors and the canines in the lower jaw). The molars were low-crowned with cusps forming low ridges (fig. 185e). The most notable feature of its dentition was the enlarged size of one pair of incisors in each jaw which may be regarded as incipient tusks (fig. 185b).

During the middle and later Cainozoic, proboscideans of varied types developed and dispersed over much of the Old and New World. They represent several separate lines which, however, developed a number of similar features such as a great increase in size, growth of pillar-like legs (fig. 179f), growth of a trunk, and above all, specialisation of the teeth (fig. 185a, c, d). One lineage only survived the Pleistocene; it includes the Ice Age mammoths as well as existing elephants.

PERISSODACTYLS. The living members of this group, sometimes referred to as the 'odd-toed' ungulates, include the horse, zebra and rhinoceros. They are a small remnant of a once very extensive group which was at its acme in late Eocene and Oligocene times (fig. 183a, tooth). Apart from the most primitive forms they have either three toes, in which case the axis of the foot passes through the enlarged middle toe (third), or one toe only.

HORSES. The descent of the horse, *Equus*, is particularly well documented by fossils and is one of the classic examples of evolution. Much of the fossil material was collected in North America where there is a fairly continuous series of Tertiary sediments. Some fossil horses are also found in Europe and South America.

The ancestral horse, *Hyracotherium*, occurred widely in the early Tertiary in North America and Europe; in Britain it is recorded from the London Clay. It was a small fox-sized animal with a small, relatively simple brain. It had four toes on each foreleg (fig. 186c), and three on each hind leg, each toe ending in a hoof. Its teeth were of simple form, including small chisel-shaped incisors, small canines and low-crowned cheek teeth with blunt cusps (fig. 186h, i).

The main line of descent leads directly to the modern horse, *Equus* (fig. 186b) and related forms like the zebra, which appeared during the Pleistocene, but in the course of the Tertiary many varied sorts of 'horses' branched off the main lineage. Some of the main progressive changes which occurred may be summarised:

(i) There was an increase in size from about 30 cm in *Hyracotherium* to over 1·5 m in *Equus*.

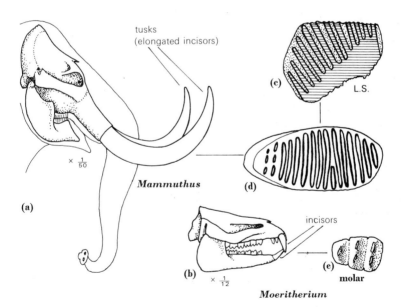

185   Elephant skulls and teeth.

a, b, skulls, in profile, of a Pleistocene mammoth (a), and of the primitive forerunner of the elephants (b) (note the shortening of the jaws and the elongation of the incisors in (a) as compared with (b)). c, a simplified section and d, e, crown view of molars; (ornament in (c) as in fig. 180).

(ii) The legs became long and slender as a result of the elongation of certain bones, while others were reduced or lost. In the foot, the middle digit was enlarged while the lateral digits were reduced and finally disappeared (fig. 186c–f).

(iii) The braincase indicates a great increase in size and complexity of the brain (fig. 186g).

(iv) The face in front of the eyes is lengthened to accommodate the increased size of the teeth.

(v) A marked gap developed between the incisors and the cheek teeth (fig. 186b); the latter became high crowned, and the pattern of ridges increasingly complex (fig. 186j, k).

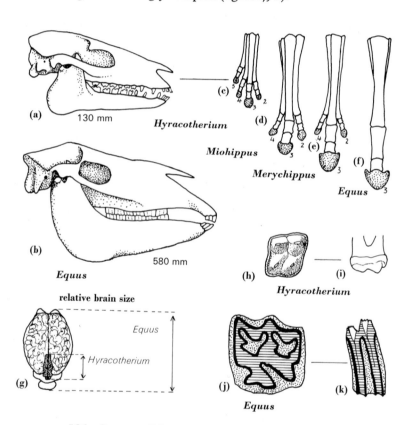

186   Some modifications of the skeleton in horses.

a, b, skulls of the earliest horse (a), and of a Pleistocene horse (b). c–f, changes in the right fore-foot in a series of horses from c, the Lower Eocene (four toes functional); d, the Oligocene (three toes functional); e, the Miocene (lateral toes reduced); f, Pleistocene and Recent (one toe). g, comparison of relative brain size. h–k, upper molars; (h, crown view; i, side view; j, transverse section; k, longitudinal section; ornament as in fig. 180).

The evolution of the horse is interpreted as a response to a change-over in its environment from a forest-type area where the early horses browsed and where the ground was relatively soft, to an open grassland type where the later forms grazed over a wide expanse of firmer ground. This conclusion takes into account evidence from other sources, including the widespread appearance in the Miocene rocks of the remains of grass seeds.

ARTIODACTYLS. The artiodactyls, including pigs, deer and cattle, are the dominant ungulates today although, coincident with the recent expansion of man, there has been a decline in their numbers. They appeared in the Eocene, and in the later Tertiary had largely displaced the perissodactyls. In contrast to the latter they have an even number of toes, two or four, with the axis of the foot running between the third and fourth toe. Forms like the pig have relatively primitive teeth, but more typically the teeth are fully modified for a herbivorous diet.

19
Plants

Plants typically possess CHLOROPHYLL, a green pigment, which absorbs energy from sunlight and thus makes possible the synthesis of carbohydrates from CO_2 and water, a process known as PHOTO-SYNTHESIS. The carbohydrates, along with simple inorganic materials, are then built up into the more complex substances from which the plant protoplasm and tissues are formed. Certain plants, like fungi, lack chlorophyll and depend for their food on other organisms, alive or dead. Most plants are anchored to the soil or, if aquatic, are non-motile although a number of active unicellular organisms are classed as plants because they contain chlorophyll and CELLULOSE. Cellulose, a complex carbohydrate, is the basic constituent of the cell wall in plants. In higher (more complex) plants the cell walls in some tissues (VASCULAR TISSUES) are thickened, consisting of a mixture of cellulose and LIGNIN (a complex aromatic compound) which makes them strong and rigid. In the higher plants, too, the outer surface is covered by an almost unbroken layer, the CUTICLE, consisting of an impervious waxy substance, CUTIN, which reduces water loss.

Preservation of plant remains

Dead plant tissue decays rapidly under normal subaerial conditions and, since in general there is no mineral matter to strengthen the structure, fossil plants occur rather sporadically and tend to be poorly preserved. In circumstances, however, which exclude oxygen and aerobic bacteria, plant material may be well preserved and in some cases, as in the Rhynie Chert, very fine details may be found.

The accumulation of plant material in stagnant, oxygen-deficient water is quite favourable to preservation and such conditions occurred extensively throughout the Northern Hemisphere during parts of the Carboniferous. The coals formed then consist of partly carbonised vegetable matter, some of which was macerated so that cuticles and spore

cases are the only identifiable structures, whereas other material preserved entire still shows under the microscope the structures of the
original woody tissue. In nodules (or COAL BALLS) occurring in some
coal seams, the plant tissues are mineralised, mainly by calcite, and show
cell structures. Some of the most perfectly preserved plants are those in
which the tissue has been replaced by silica. The Rhynie Chert (Middle
Old Red Sandstone, Aberdeenshire), an outstanding example of this
mode of preservation, represents a silicified peat containing the remains
of small animals (e.g. mites, insects) as well as plants in which the cell
structure can be distinguished in thin section (fig. 193).

Plant fossils are usually fragments of separate parts, such as roots,
stems, leaves or seeds. Complete specimens showing the connection
between the various parts are uncommon, and fossils have often been
assigned to 'form' genera which subsequent discoveries have shown to
be related: e.g. *Lepidodendron* refers to the trunk of a fossil lycopod;
Stigmaria to roots; *Lepidophylloides* to leaves; and *Lepidostrobus* to
cones.

CLASSIFICATION. Plants are classified according to details of their
structure, and their mode of reproduction. With advances in knowledge, new names have been introduced for the main divisions of plants,
but many of the categories have 'common' names which are familiar
through long usage and are, therefore, convenient to use here.

THALLOPHYTES

The simplest plants, the algae, bacteria and fungi, often grouped
together as THALLOPHYTES, consist of soft tissue showing no differentia-
tion into roots, stem or leaves. They range from microscopic one-celled
forms to more complex many-celled plants which are mainly filamentous
or have a flattened body. They may reproduce asexually by single cells,
SPORES, which develop directly into new plants, but they may also
reproduce sexually, producing gametes which conjugate to form a
zygote.

Algae

Algae are essentially aquatic plants which absorb their nutrients and
CO_2 from the surrounding water, and include microscopic one-celled
forms like diatoms and coccospheres and also the large seaweeds. As
fossils, only marine forms need be mentioned. These include benthonic
seaweeds and also planktonic forms, the PHYTOPLANKTON, all of

187   A stromatolite; polished section of Islay Limestone, Dalradian Series,
Pre-Cambrian, Argyll (×2).

which are of great importance in marine ecology, since they provide
food directly or indirectly, for all aquatic animals, and, by their photosynthesis, oxygenate their environment. Most leave little or no fossil
trace. However, a number of forms which secrete calcium carbonate, or
more rarely silica, may be preserved as fossils, and their record dates
back far into Pre-Cambrian times.

Benthonic algae

Since these seaweeds require light for photosynthesis, they occur in
relatively shallow water within the photic zone, ranging down to about
200 m in tropical waters and about 100 m in higher latitudes. They are
commonest in tropical seas, and some play an important role in reef-
building.

Some algae may stabilise the soft sediments in which they live by the
felted growth of their filaments. Where the deposits are limey, roughly
bun-shaped structures may form. Similar laminated structures, stromatolites, are common in some Pre-Cambrian and Palaeozoic rocks.

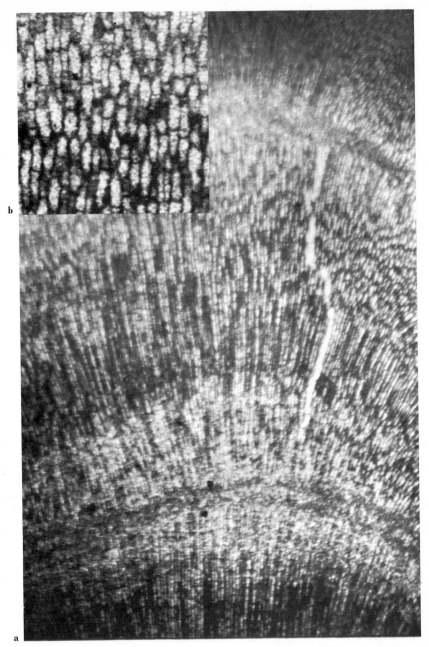

188   Thin sections of a calcareous alga; *Solenopora*, Bathonian, Jurassic, Glos.

a, part of the alga showing banding (×16); b, cell detail (×60).

189   A coccolith, Miocene, Jamaica ($\times$ 13,500).
(Electron micrograph.)

Pre-Cambrian examples include limestones in Rhodesia which are possibly over 2600 million years old, and the stromatolitic limestones of the Belt Series in the Eastern Rockies, which are of much later date. The latter take the form of corrugated, laminated sheets without cell structures occurring in the limestones. In Britain, 'stromatolites' occur in limestones in the Dalradian Series (fig. 187) and in the Durness Limestone.

CALCAREOUS ALGAE. Several different types of algae, collectively referred to as the calcareous algae, secrete $CaCO_3$ and are readily preserved. Their fossils are common at many horizons especially in reef deposits. Common examples include *Lithothamnion* (Cretaceous to Recent), *Solenopora* (Ordovician to Jurassic) and *Girvanella* (Ordovician to Jurassic). These form nodular masses and, in thin section, show a banded structure with fine branching tubes which may be arranged radially (fig. 188) or, as in *Girvanella*, in a tangled mass.

Phytoplankton

CALCAREOUS FORMS. COCCOSPHERES are green algae with minute calcareous structures, COCCOLITHS, embedded in their cell walls. They are flagellates and swim freely in the surface waters (top 100 m) of the ocean, occurring in greatest abundance in temperate zones. Coccoliths (fig. 189) are regular oval structures made up of calcite platelets arranged in elaborate patterns; their detailed structure has been made clear only since their examination under an electron microscope. Recent studies of Mesozoic and Tertiary rocks have shown an abundance of coccoliths in many horizons. In particular, the Chalk owes much of its lithological character to an exceptional abundance of coccoliths. Some species have a restricted range and are thus of value for correlation.

SILICEOUS FORMS. The cell wall in DIATOMS (Jurassic to Recent) is impregnated with silica and forms a skeleton of two overlapping valves, with a finely reticulate structure (fig. 190). Diatoms occur both in fresh water and in the sea, living near the surface and being most abundant in colder waters. Deposits rich in diatoms are found in fresh-water lakes in high latitudes and in the sea in polar regions. Diatomaceous earths of Pleistocene age occur in more northern parts of the British Isles, e.g. in the Lake District. Marine rocks, diatomites, of Miocene age are found in parts of the United States.

OIL-SECRETING ALGAE. A number of algae, e.g. some unicellular blue-green algae, synthesise oil and their remains may accumulate to form sapropelic deposits including oil shales. Masses of globular yellow bodies, believed to be unicellular algae, may be seen in thin sections of the Scottish Oil Shales and torbanites.

Acritarchs

Acritarchs are unicellular microfossils of uncertain affinity. They have a hollow test of an organic nature which may be spherical, ellipsoidal or polygonal in shape. The surface may be smooth or granulated and may have spinose projections (fig. 191).

Acritarchs are included here along with algae since they most nearly

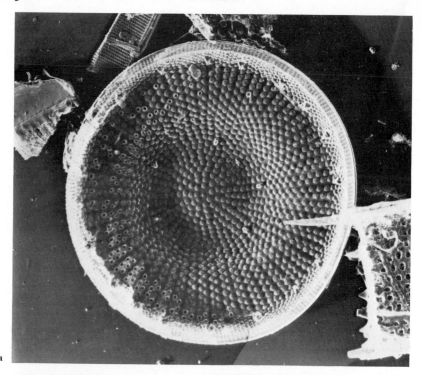

a

b

190   Diatoms.

a, from diatomite, Neolithic, Kenya ($\times$2550). b, from Loch Droma, Quaternary ($\times$11,500).
(Electron micrographs.)

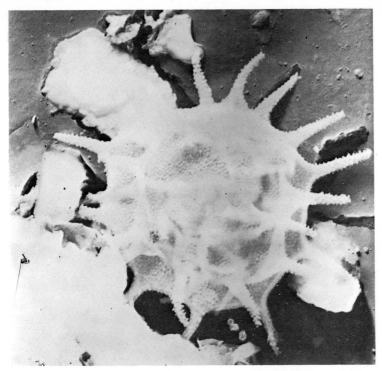

191   An acritarch, Upper Eocene ($\times$ 12,000).
(Electron micrograph.)

resemble cysts or vegetative stages in the life cycle of certain algae
(Dinophyaceae); but there seems some possibility that a number will
eventually be identified otherwise, as, for instance, spores or egg cases.
They are of some stratigraphic value, especially in marine rocks.

Bacteria

Bacteria are ubiquitous microscopic organisms which occur as single
cells or aggregates of cells and lack chlorophyll. The cells may be as
little as 0.5μ and are spherical or rod-shaped. They multiply by simple
fission and may form spores which are highly resistant to temperature
and desiccation.

Most bacteria live on organic matter, alive or dead, but some (of
considerable interest to the geologist) derive their energy from chemical
reactions of inorganic material, e.g. the sulphur and iron bacteria. Also,
a number are ANAEROBIC, living in environments lacking oxygen; for
instance, some forms liberate methane by breaking down organic

**192   A psilophyte; a primitive vascular plant from the Lower Old Red
Sandstone (Senni Beds), South Wales (×1·2).**

matter, and others liberate hydrogen sulphide by decomposing organic
matter or by reducing sulphates in water.

Bacteria lack hard parts and evidence of their existence in earlier
geological times is largely circumstantial. Since at the present time they
play a major role in the breakdown of organic matter, presumably they
have done so in the past; for instance, methane is a commonly occurring
natural gas. Anaerobic bacteria are invoked to explain occurrences of
euxinic shales (i.e. shales formed in a non-oxygenated environment),
and the activities of iron bacteria to account, at least in part, for the
formation of some iron deposits.

Pre-Cambrian to Recent.

Fungi

Fungi, such as moulds and toadstools, are unicellular or consist of
filaments (HYPHAE). They lack chlorophyll and feed on organic matter
through which the hyphae ramify. They have no resistant tissue and are
only exceptionally preserved as fossils, as for example in the silicified
plant tissues in the Rhynie Chert.

Devonian to Recent.

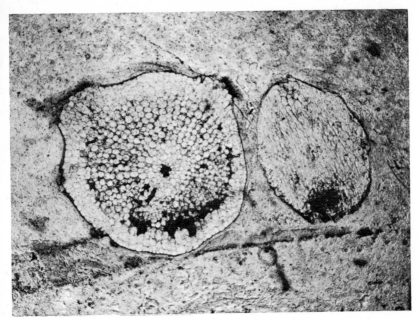

193 Cell structure in Rhynie Chert, Middle Old Red Sandstone, Aberdeen-shire ($\times$35). A transverse section of *Rhynia* is included; the small dark area, in the centre of this, represents woody tissue.

MOSSES

Mosses (or bryophytes) are the simplest land plants. They lack the vascular tissue (water- and food-conducting strands) which strengthens higher plants, and they show a primitive life cycle involving a distinct ALTERNATION of sexual and asexual generations, which is dependent on water for its completion. Thus they occur characteristically in damp places.

The moss plant with which most people are familiar represents the sexual GAMETOPHYTE phase and bears male and female gametes; these unite to form a zygote from which the asexual SPOROPHYTE phase develops. The gametophyte consists of aerial shoots with leafy structures and is anchored in the soil by hair-like processes, RHIZOIDS, which absorb water and nutrients. The male gamete is motile and can fertilise the female gamete only if water is present; dew is adequate. The sporophyte, a stalked capsule, remains attached to the gametophyte. Numerous spores develop in it and, when ripe, are dispersed by wind; they germinate in damp soil and grow into a gametophyte.

Mosses appear first in the Upper Devonian. They are, generally, rare as fossils, except in the Quaternary where they are very abundant.

VASCULAR PLANTS

In higher land plants, the body is differentiated into aerial shoots, or STEMS, with LEAVES in which photosynthesis occurs, and underground anchoring ROOTS which absorb water and nutrients. Water and food move through the plant along special conducting VASCULAR TISSUES in which the cell walls are woody, being thickened with LIGNIN and CELLULOSE (p. 294). The woody vascular tissue also gives mechanical support, so that these plants are not restricted in size as are the more primitive mosses, and some, including the forest trees which grow annual increments of secondary wood, may reach great size. Water is conserved in the vascular plants by an impervious waxy layer, the CUTICLE (p. 294), covering the surface of stem and leaves. Interchange of gases (CO_2 and O_2) with the atmosphere takes place through breathing pores, STOMATA, in the cuticle, and these also control the escape of water vapour from the plant.

As in the mosses (p. 303), the life cycle of vascular plants involves an alternation of generations, but, in contrast to mosses the sporophyte phase is dominant and the gametophyte is inconspicuous. There are two major subdivisions, the spore-bearing PTERIDOPHYTES and the seed-bearing GYMNOSPERMS and ANGIOSPERMS.

Pteridophytes

Pteridophytes embrace the spore-bearing vascular plants. Those with a fossil record include the psilophytes, clubmosses (or lycopods), the horsetails (Sphenopsida) and ferns (Pteropsida). They are for the most part small herbaceous plants, having no growth of secondary wood. Exceptional fossil forms, however, grew secondary tissue and reached great size, e.g. the lycopods found in the Coal Measures.

The pteridophytes reproduce by spores which are borne on spore-bearing leaves, or SPOROPHYLLS (fig. 194g). Most have spores of one size only, but some, like the modern lycopod *Selaginella* (and certain fossil forms) have large MEGASPORES and small MICROSPORES. The megaspores germinate to form the female gametophytes which produce egg cells, and the microspores develop into the male gametophytes. The latter bear motile gametes which require the presence of water for union with the egg cells. In some species the megaspore remains attached to its

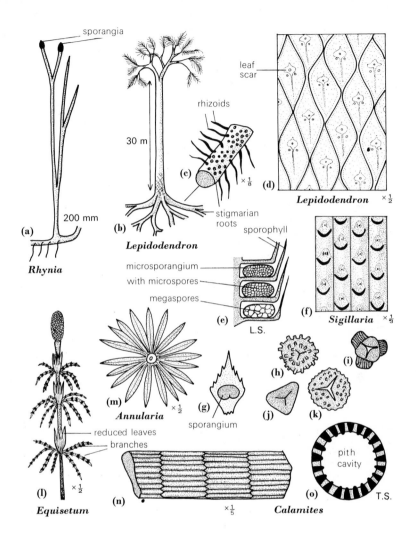

sporangia

leaf
scar

rhizoids

30 m

(c) ×⅛

(d)

Lepidodendron ×½

200 mm

(a)

(b)

stigmarian
roots

sporophyll

Lepidodendron

Rhynia

microsporangium

with microspores

megaspores

(e)

L.S.

(f)

Sigillaria ×⅓

(m)

Annularia ×½

(g)

sporangium

(h)

(i)

(j)

(k)

reduced leaves

branches

(l) ×½

Equisetum

(n)

×⅕

Calamites

pith
cavity

(o)

T.S.

194 Fossil plants.

a, a psilophyte from the Middle Old Red Sandstone. b, a restoration of a Carboniferous tree-like
lycopod. c, fragment of a stigmarian root. d, f, Carboniferous lycopods with a diagonal (d), and
a vertical (f), arrangement of the leaf scars. e, a longitudinal section of part of a lycopod cone.
g, a sporophyll of a modern lycopod. h–k, Coal Measure pteridophyte microspores, much
enlarged. l, a modern horsetail. m–o, parts of Coal Measure horsetails; m, leaves; n, internal
mould of a stem; o, a diagrammatic transverse section of a stem, to show the wedges of woody
tissue projecting into the pith cavity.

parent sporophyll, where it develops into the female gametophyte, so that fertilisation of the egg cell takes place on the sporophyll. This marks a stage towards the condition found in seed plants.

Psilophytes

Psilophytes are the most primitive spore-bearing plants and they are also the earliest of the vascular plants to appear as fossils, occurring in the Upper Silurian. They formed an important part of the Devonian flora, but are not found in later rocks. Two living genera, found in warm climates, show a similar organisation.

Psilophytes are recorded in Britain from several horizons in the Devonian, including the Downtonian (fig. 192). The best known are those found in the Rhynie Chert (presumed Middle Old Red Sandstone) in which several genera are very finely preserved in silica (p. 10). One of these, *Rhynia* (fig. 194a) consisted of leafless stems which forked dichotomously, arising from a horizontal creeping stem held in the soil by rhizoids. Vascular tissue can be distinguished in thin section (fig. 193) and also cuticle with stomata (p. 304). There were spore-bearing organs, SPORANGIA, at the tips of some of the stems.

Lycopods

Living lycopods are small herbaceous plants with stems which sprawl over the ground and from which erect branches rise at intervals. They have a dense cover of small sharp leaves, spirally arranged and each with a single vein (i.e. a strand of vascular tissue). Some leaves are sporophylls, bearing kidney-shaped sporangia (spore-producing organs, fig. 194g). The sporophylls may be clustered into loose cones.

The earliest lycopod, *Baragwanathia*, is found in Australia in association with a Lower Devonian (Gedinnian) species of *Monograptus*. A variety of lycopods are found in the later Devonian and the group reached its acme in the Carboniferous, being especially prominent in the Coal Measures.

Lepidodendron (Coal Measures, fig. 194b) is one of the best known fossil lycopods. It reached tree-like proportions, trunks about 30 m in height with a diameter of over 1 m having been recorded. The trunk had some secondary wood, but the main increase in girth was due to a thick outer layer of non-woody cells. The roots grew out almost horizontally, forking dichotomously (fig. 194b), with round scars on the surface marking the former position of rhizoids (fig. 194c). Such roots, typical in fossil lycopods, are referred to the form genus *Stigmaria*. The trunk was profusely branched at the top and the stems were densely covered with elongate strap-like leaves (form genus *Lepidophylloides*) which, as they were shed, left prominent spirally arranged leaf scars with a characteristic rhomboid shape (fig. 194d). A related genus, *Sigillaria* (Coal Measures), had squarish leaf scars arranged in vertical rows (fig. 194f). The sporophylls were clustered in cones (form genus *Lepidostrobus*) at the ends of small branches. Both megaspores and microspores were produced in vast quantities, in some species in the same cone (fig. 194e) but in others, in separate cones. The spores have an outer cover (exine) which is resistant to decay and may be abundant in some coals, especially in durain layers from which they may be extracted by chemical methods (fig. 194h–k). Assemblages of such spores have been used empirically for correlation of coal seams over limited areas. The tree-like (arborescent) lycopods disappeared at the end of the Palaeozoic.

Horsetails

The small herbaceous *Equisetum* is the only surviving genus of horsetails (fig. 194l). It has an underground stem from which arise jointed aerial stems with much reduced leaves. Whorls of slender branches diverge from the main stem at each joint (node) and some of these bear sporangia in groups at their tips.

Horsetails appeared in the Devonian and were abundant in the Coal Measures, declining in numbers rapidly after the Palaeozoic. The commonest fossil is *Calamites*, an arborescent form which reached a height of about 27 m. It had a much-branched stem in which, unlike *Equisetum*, there was a development of secondary wood. The axial part of the stem contained soft tissue (PITH) which decayed rapidly, leaving a cavity. Internal moulds of this pith cavity are common and may be round, or flattened, according to the type of infilling sediment. The moulds show the transverse markings of the leaf nodes and longitudinal furrows and ridges (fig. 194n). The furrows are impressions left by radial wedges of wood projecting into the pith cavity (fig. 194 o). The leaves of some species of *Calamites* are assigned to the form genus *Annularia* (figs. 194m and 195); they are small and lance-shaped, and are arranged in rosette-like whorls. The sporangia were borne in compact cones at the tips of the branches.

Ferns (pteropsids)

A typical fern has an underground stem, from which roots grow down into the soil, and, above ground, grow large leaves, the FRONDS, which

195   Leaf whorls of an Upper Carboniferous horsetail; *Annularia* (×1·3).

196 A scrambling seed-fern; *Calymmatotheca*, Coal Measures, Upper Carboniferous, Kilwinning, Ayrshire (×1·3).

typically are divided and subdivided into smaller leaflets, PINNAE. The veins in each pinna are forked. Sporangia lie on the underside of the pinnae.

Ferns were abundant in the Carboniferous, some forms reaching a height of 15 m. They remained important until the Jurassic. *Tempskya* is a Wealden 'tree-fern'. The dominant living families of ferns date from the end of the Mesozoic and are widely distributed, though occurring more commonly in tropical areas where some tree-ferns may grow to a height of 12 m.

Many fern-like fossils occurring in the Carboniferous are not true ferns, but 'SEED FERNS' (p. 312), which can be distinguished only if sporangia or seeds are present.

Seed-bearing plants

The higher vascular plants reproduce by SEEDS, i.e. embryo plants with bud, root and seed leaves covered by a protective coat. The male gamete is non-motile (except in *Gingko* and cycads) and so a film of water is not required for its union with the egg. This occurs on the plant since, as was noted in some lycopods (p. 304), the megaspore is retained in its megasporangium. The latter is covered by extra layers of tissue and is called an OVULE. The microspores are referred to as POLLEN GRAINS, and they are transferred to the ovule either by wind or, as in some of the higher plants, by a more complicated process, for instance, by insects. It may be noted that the walls of the pollen grains are sculptured with distinctive markings which are characteristic of different species. Thus pollen grains extracted from sedimentary rocks are a valuable means of identifying past floras.

There are two major subdivisions of seed plants, the GYMNOSPERMS, which are the more primitive, and the ANGIOSPERMS. They differ in a number of ways, including the nature of the reproductive organs which form CONES in gymnosperms, and FLOWERS in angiosperms. A further distinction is the state of the ovule at the time the pollen grain is transferred to it. In gymnosperms, the ovule is exposed on the surface of the sporophyll; in angiosperms, it is encased by the sporophyll, which forms a protective ovary.

Gymnosperms

Gymnosperms include the SEED FERNS (pteridosperms), CYCADO-PHYTES, GINKGOS, and CONIFERS. They are woody plants which increase in girth by annual growth of secondary wood.

The basic pattern of reproduction is similar throughout the gymnosperms, though in each group there is a characteristic variation in detail. The sporophylls, bearing sporangia, are usually aggregated in cones. The megaspore, retained in its OVULE (megasporangium, p. 310) grows into a female gametophyte in which an egg cell develops. The POLLEN GRAINS (microspores, p. 310), each containing a miniature male gametophyte, are shed in enormous numbers into the air and are windborne to the ovules. When a pollen grain makes contact with an ovule male gametes are released, one of which fertilises the egg cell. The resultant zygote develops, within the ovule, into an embryo sporophyte plant, the SEED. The seed may remain dormant for a period until conditions are suitable for its germination.

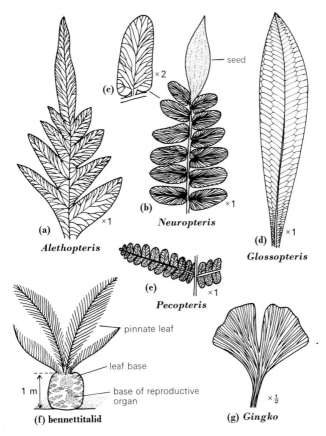

197   Fossil plants.

a–e, seed-ferns. f, restoration of a cycadophyte; most of the leaves have been omitted. g, leaf of a modern *Gingko*.

Seed ferns

The SEED FERNS, or pteridosperms, are extinct plants which had much-divided, fern-like, leaves attached to a slender aerial stem (fig. 196). They reproduced by seeds borne on the leaves, and the pollen-bearing sporangia were arranged in separate clusters on the fronds. Reference has been made (p. 310) to the fact that certain fossils with fern-like leaves can only be identified as seed or true ferns if the pinnae are fertile. For instance, in the case of *Pecopteris* (Carboniferous, fig. 197e), some species had seeds and were seed ferns, but other species were true ferns with sporangia on the underside of the pinnae.

Seed ferns first appeared in the Upper Devonian and were at their acme in the later Palaeozoic when genera like *Neuropteris* (fig. 197b) and *Alethopteris* (fig. 197a) formed a notable part of the Coal Measure flora; subsequently they gradually declined to extinction in the Jurassic.

The genera mentioned occurred widely, in Coal Measure times, in the Northern Hemisphere. A flora of markedly contrasting type flourished, from the late Carboniferous through Permian and Triassic times, in the southern continent of Gondwanaland (i.e. peninsular India, Africa, Australia, Antarctica and South America). It is named after the most characteristic member of the assemblage, *Glossopteris* (fig. 197d). In this form seed-like structures have been found attached to the tongue-shaped leaves, and it is, accordingly, generally classified as a seed fern.

Cycadophytes

The cycadophytes have a rough, columnar stem covered by the leaf bases of the growth of previous seasons, and crowned by pinnate leaves of tough, leathery texture. They include two groups of plants which have an important fossil record in the Mesozoic. One, the BENNETTI-TALES, is extinct; and the other, the CYCADS, still survives. Superficially alike, they differ in structural details, and in the reproductive organs. The Bennettitales had a short columnar trunk with hermaphrodite cones, of almost flower-like appearance, growing between leaf bases (fig. 197f). The ovules were borne on a central conical structure which was surrounded by pinnate structures carrying the pollen; the whole cone was protected on the outside by hairy leaf-like structures.

The Bennettitales (Trias to Upper Cretaceous) had a world-wide distribution during the Jurassic and Lower Cretaceous. In Britain they occur, for instance, in the Jurassic Deltaic Series in Yorkshire (fig. 198) and also in the Purbeck 'dirt' beds where well-preserved trunks are found.

198   Leaf of a bennettitalid; *Williamsonia*, Lower Deltaic Series, Jurassic, Yorkshire (×1).

A few cycads survive today in tropical countries. They differ from the previous plants in the form of their cones which are compact, barrel-shaped structures borne at the end of the stem, and also in that individual plants bear either ovules or pollen. Their range is from the Trias to the present day.

Ginkgos

A single surviving species of *Ginkgo*, the maidenhair tree from Southern China, represents a race of plants which first appeared in the Upper Palaeozoic and was very widespread in the Mesozoic. *Ginkgo* is a deciduous tree, reaching about 30 m in height and branching repeatedly. Individual plants are either male or female, the male tree carrying its sporangia in pendulous cones, and the female tree carrying pairs of ovules on short stalks. The leaves are fan-shaped with veins which branch regularly (dichotomous branching) (fig. 197g). The leaves are the parts most commonly preserved as fossils, occurring, for instance, in the Deltaic Series in Yorkshire (fig. 199), and in the Ardtun leaf beds of Tertiary age in Mull. Some of the fossil leaves (occurring in rocks which range in age back to the Jurassic) are identical with those of the living *Ginkgo biloba*, while others are deeply divided and are referred to separate genera.

Conifers

Conifers are, typically, tall trees with evergreen needle- or scale-like leaves. The ovules and pollen grains are borne in separate cones on the same individual. The ovules lie under leaf-like scales in a compact cone, and the pollen grains may have wing-like projections which aid in their dispersal by wind (fig. 14b).

Conifers are a flourishing group at the present day, occurring widely in temperate zones where they may form dense forests. They include the familiar pine and larch, *Araucaria* (monkey-puzzle tree), and the tallest of all forest trees, the redwoods (e.g. *Sequoia*, which grows to about 106 m).

Conifers appeared in the later Palaeozoic and, in the Mesozoic, they were, together with the Bennettitales and ginkgos, the dominant land plants. Fossil conifer logs are common in some parts of the Wealden Series. By late Cretaceous, however, conifers were surpassed in numbers of genera by the angiosperms.

A conifer-like tree, *Cordaites* (Carboniferous and Permian), which was abundant in the Coal Measures may conveniently be included here.

199 *Gingko* leaves, Inferior Oolite, Jurassic (×1·2).

200   Cinnamon leaf, Miocene ($\times$ 1·3).

It was tall (30 m), with a crown of branches bearing spirally arranged strap-like leaves measuring up to a metre in length.

Angiosperms

The ANGIOSPERMS, or flowering plants, are the dominant plants today. They include woody and herbaceous members which show an enormous range in structure and habit. Typically, the male and female organs occur on the same plant and, usually, in the same flower. Two classes are distinguished, MONOCOTYLEDONS which have one seed leaf, and DICOTYLEDONS which have two seed leaves.

MONOCOTYLEDONS include palms, grasses and bulbs like onions and tulips, plants which do not usually grow secondary wood. The leaves have parallel veins, and the flower parts are grouped in threes. DICOTY-LEDONS include most of the flowering plants and, also, all the forest trees (other than the conifers) and shrubs in which secondary wood is formed each season. The leaves have branching (reticulate) veining, and the flower parts are in fives or fours.

The angiosperms appeared in small numbers in the Cretaceous by the end of which they were widespread, richly diversified and had displaced the gymnosperms to become the predominant land plants. Many of the genera still survive, including such familiar examples as the holly, poplar, maple and magnolia.

Tertiary floras are abundantly preserved in many parts of the world, and since they include an increasing proportion of modern species they have provided valuable information about climatic changes. The flora of the London Clay has been extensively examined. Most of the fossils are angiosperms (mainly fruit, seeds and leaves) and some of these survive today in tropical conditions. One fossil closely resembles the fruit of the modern *Nipa* palm which grows in brackish water at sea-level in the Indo-Malay peninsula. Other fossils include the leaves of such forms as the cinnamon (fig. 200) and fig which today grow in subtropical areas.

The occurrence of husks of grass seed, in early Miocene deposits in parts of America (e.g. Nebraska), indicates the development of open prairie grasslands, replacing forest conditions. This development is noteworthy in connection with the expansion, at the same time, of various herbivorous mammals.

Reference has already been made (p. 28) to the value of fossil plants in deciphering details of climatic changes during and since the Pleistocene. Much of the evidence is based on an analysis of pollen grains.

TECHNICAL TERMS

CONE a reproductive structure, found in conifers and some pteridosperms, and consisting of sporophylls (*q.v.*) arranged compactly on a central axis.

GAMETE a reproductive cell which unites with another gamete to form a zygote, from which a new individual grows.

GAMETOPHYTE the phase of a plant which bears the sex cells, or gametes.

MEGASPORANGIUM the sporangium in which the megaspore (*q.v.*) is formed.

MEGASPORE the larger of the two sizes of asexual spores produced by some pteridophytes, which germinate to form a female gametophyte; it is also represented in seed plants, being contained in the ovule.

MICROSPORANGIUM the sporangium in which the microspores are formed.

MICROSPORES the smaller of the two sizes of asexual spores found in some pteridophytes; they germinate to form the male gametophytes. In seed plants, microspores are known as pollen grains (*q.v.*).

OVULE the structure, in seed plants, which contains the megaspore; it is equivalent to the megasporangium of pteridophytes.

PINNA a leaflet.

PINNATE refers to a compound leaf consisting of a series of pinnae arranged on each side of a common stem.

POLLEN GRAINS structures, found in seed plants, which are the equivalents of microspores in pteridophytes. Each pollen grain contains a microscopic male gametophyte. It has a resistant outer coat bearing distinctive sculpturing.

RHIZOID a hair-like root structure.

RHIZOME an underground stem.

SPORANGIA structures within which asexual spores are formed.

SPOROPHYLL a leaf which bears sporangia (*q.v.*). It may resemble the ordinary leaves of the plant, or, as in seed plants, it may be modified and unlike the other leaves.

SPOROPHYTE the phase of a plant which bears asexual spores.

VASCULAR TISSUE the plant tissue, including the wood, along which water and food are conducted in the plant, and which also supports the plant.

VEIN a strand of vascular tissue.

ZYGOTE the fertilised cell which results from the union of a male and a female gamete.

20

Epilogue: evolution and the origin of life

We have seen that the study of fossils is inextricably bound up with a wider field of study, the geological and biological sciences in general. The question now arises, 'What contribution does palaeontology make to this broader field of study?' The answer to this is that it provides the time dimension. Whereas the study of biology is necessarily restricted to present-day organisms, palaeontology is concerned with the distribution of organisms in time, over a period of 600 million years or more.

The fact that the fossil record is necessarily incomplete should not obscure another fact, that an enormous amount of information about the history of life is now available. The over-all picture which emerges from this study of the history of life is a dual one. One is impressed, on the one hand, by the great diversity of organisms which have appeared; and, on the other hand, by the continuity of the few basic life patterns within which this diversity has been contained.

During the course of geological time the numerical importance of the different groups of organisms has varied considerably, and some of this variation is shown in fig. 201. The most striking feature of this diagram is the appearance (at the beginning of the Cambrian and within a period of a few million years) of invertebrates with the ability to build a skeleton. The first of these, the trilobites and brachiopods were followed by the sponges, coelenterates, gastropods and echinoderms, in the course of the Lower Cambrian. This was one of the major events in the history of life, but we must leave it to the biochemist to discover how organisms, which were, of course, already producing complex substances with which to build their bodies, began also to make mineral compounds, such as carbonate and phosphate of calcium, to strengthen their soft parts.

Another interesting feature of the diagram is the way in which some groups, e.g. the bivalves and gastropods, with a very long geological record, have increased in numbers quite steadily over a period of 500 to 600 million years. These seem to be groups which have been able to adapt themselves to changing conditions with very great efficiency. On the other hand, the brachiopods reached a numerical peak during the Palaeozoic, but have declined in importance since then. This could be due to competition from other groups with a similar mode of life and diet, for example, the increasingly successful bivalves.

Some groups, of course, have become extinct, notably the trilobites, which were the most important forms during the Cambrian and rose to their numerical peak in the Ordovician, but which declined rapidly in importance thereafter. The graptolite record shows the same pattern.

Another interesting feature shown by the diagram is the way in which some major groups of invertebrates appeared much later in geological time. The ammonoids, for instance, appeared at the start of the Devonian; and an entirely new group of corals, the scleractinians, made their debut in the Trias.

When we turn to the vertebrates, we see a series of major groups which appeared one after the other, over a very long period of time. The first vertebrates, the primitive fish, were followed by the amphibia, by the reptiles and by the mammals. Many lineages, within these groups, subsequently became extinct and all, except the mammals and fish, are of less importance now than in the past.

All these changes can be summed up in the word 'evolution', the hypothesis which successfully links together all the palaeontological and biological data about life. The fossil record indicates that evolution has taken place on at least two different scales. On the one hand, we have a series of small, gradual changes, such as occurred in *Micraster*, during the Upper Cretaceous (p. 123). These changes were quite significant, but so slight that they did not alter the general character of the genus, which remained *Micraster* from the beginning to the end. The time over which these changes took place was of the order of 10 million years. On the other hand, we have evolutionary change which resulted in the appearance of the first amphibia from the fish, or the first reptile from the amphibia; relatively large changes, which apparently took place relatively quickly and without a long series of transitional forms. It appears, therefore, that there have been periods in geological time when major evolutionary steps have taken place quickly, and other periods when the process operated much more slowly.

A detailed explanation of the mechanism of evolution is, of course, a matter for the biologist and the biochemist, but any hypothesis put

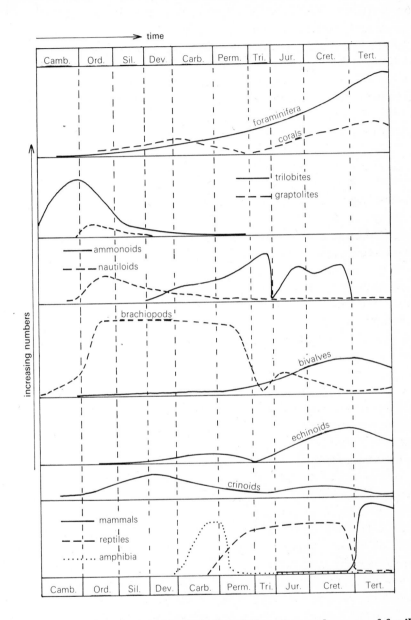

201 Diagram to show the numerical variation of several groups of fossils during geological time.

forward must be tested against the palaeontological evidence. The most likely explanation is one based on two factors, gene mutation and natural selection. A mutation can be defined as a 'sudden alteration in the composition of a gene' (i.e. one of the units from which chromosomes are formed and which determine inherited characteristics) as a result of which some character is permanently altered in appearance. Mutations are known to occur quite commonly at the present day and presumably have done so throughout the history of life. It follows, therefore, that a very large, total number of mutations must have occurred in the past. The overwhelming majority, however, did not perpetuate themselves due to the operation of the second factor, natural selection. Any new mutant, which is changed in such a way that it is at even a slight disadvantage with its competitors, will not survive. On the other hand, the very rare mutant which is adapted to exploit its environment more successfully than its competitors will not only survive but will, by pressure of competition, tend to eliminate its competitors and will become numerically dominant.

The evolution of *Micraster* is quite convincingly explained on this basis. There was a large population of this genus, living in fairly stable environmental conditions, for a period of the order of 10 million years. Over that period, many millions of individual breeding cycles occurred with the chance of very many mutations appearing. Any mutation which increased the efficiency of *Micraster* in its burrowing mode of life was perpetuated. The others vanished.

Natural selection operates through the environment, which is in part organic, consisting of the existing fauna and flora; and in part physical, including such factors as climate and topography, which have an overall influence on distribution of life.

There is evidence that these physical factors have varied considerably during geological time. There have been periods of mountain building, of great volcanic activity and of climatic change. These events, however, have been localised or gradual or both; and it is doubtful if they have directly influenced the course of evolution. Changes in the composition of the atmosphere, however, about which little or nothing is yet known, could have varied the radiation penetrating to the earth's surface, with direct implications for living organisms.

It is not known, with any certainty, why groups of animals have become extinct in the past. But the explanation may lie partly in competitive pressure from newly emerging groups. It seems quite possible that the appearance of an unusually successful new group might, by

202   *Charnia masoni*, a Pre-Cambrian fossil first found by a schoolboy
in Charnwood Forest, Leics. (×0·7).

intense competition for food, or by direct predation, result in the elimination of some other group. Thus, the trilobites, one of the first skeletal invertebrates to appear, must have had few or no competitors initially. But their decline towards extinction coincides with the appearance of the large orthoceratids, the early fish and the eurypterids which may have been direct predators.

In any consideration of organic evolution, an inevitable question is, 'How did life begin?' The answer, of course, lies far back in Pre-Cambrian times, and the evidence is largely circumstantial, for few authentic fossils have been recorded, despite active search over many years.

Some of the fossils found in Pre-Cambrian rocks cannot be assigned with certainty to a particular group of organisms; an example is *Charnia* (fig. 202) from the Pre-Cambrian in Charnwood Forest. Others, however, are more definite and include impressions of jellyfish, and an annelid worm from South Australia; these are of late Pre-Cambrian age. The evidence of plants in Pre-Cambrian times is quite convincing and takes the record of life back in time some 2600 million years, this being the age of limestones in Rhodesia which contain calcareous algae. More circumstantial evidence of life in the Pre-Cambrian is provided by the widespread occurrence of carbonaceous rocks. Only organisms can 'fix' carbon in any quantity, and analysis of the $^{12}C/^{13}C$ ratio, in one particular instance, accords with that of organic carbon. Further, traces of amino-acids (the structural units from which proteins are built up) have been reported from Pre-Cambrian rocks.

Even so, the actual origin of life is not demonstrated by any of this evidence. For this we must turn to other sources, in particular to the chemist and biochemist.

Experiments in the laboratory have shown that mixtures of substances like methane, ammonia, hydrogen and water vapour, when subjected to an electrical discharge in the absence of free oxygen, produce quite complex 'organic' molecules, including amino-acids. Now it is probable that the primitive atmosphere of the earth differed chemically from that of the present day; that it consisted of substances like methane, ammonia, water vapour, and hydrogen; and that it lacked both carbon dioxide and free oxygen.

An atmosphere of that composition could not support life, as we know it. But it contained the raw materials from which the organic compounds might have been built. It is far from clear, however, how the formation of amino-acids, protein and other complex molecules may have developed into creatures able to reproduce themselves. Organised life, as we know it, could only have begun when the atmosphere contained some oxygen and carbon dioxide, and it seems likely that the oxygen came from the photosynthesis of plants. The first appearance of these, and of the oxygen which they produced must, therefore, have been one of the most crucial events in earth history.

SUGGESTIONS FOR FURTHER READING AND REFERENCE

Handbooks published by the British Museum (Natural History) London:

British Cainozoic Fossils, British Mesozoic Fossils and *British Palaeozoic Fossils* contain illustrations of many British fossils and give their geological range.

Fossil Amphibians and Reptiles, Dinosaurs, Fossil Birds and *History of the Primates* give a general account of these groups and make reference to specimens on exhibit in the Museum; illustrated with line drawings and photographs of restorations and reconstructions.

British Regional Geology (Institute of Geological Sciences) H.M.S.O.: a series of handbooks on the geology of Britain which are a useful source for the regional occurrence of fossils in Britain.

Yonge, C. M., *The Sea Shore*, New Naturalist Series, Collins, London, 1966. An account of the natural history and ecology of the shores around Britain.

Brouwer, A., *General Palaeontology*, Oliver and Boyd, 1967. Deals with general aspects of palaeontology including ecology and evolution.

Moore, R. C. (Editor), *Treatise on Invertebrate Palaeontology*, Kansas University Press, 1953—. Published in volumes each dealing with a separate group on a systematic and detailed basis.

La Porte, L. F., *Ancient Environments*, Prentice-Hall, Inc., 1968.

McAlister, A. L., *The History of Life*, Prentice-Hall, Inc., 1968.

Raup, D. M., and Stanley, S. M., *Principles of Paleontology*, W. H. Freeman, and Co., San Francisco, 1971.

Romer, A. S., *The Vertebrate Story*, University of Chicago Press, 1959. A semi-popular account of the evolution of the vertebrates.

Walton, J., *An Introduction to the Study of Fossil Plants*, A. and C. Black, London, 1940.

STRATIGRAPHIC POSITION OF BRITISH ROCK FORMATIONS AND STAGES MENTIONED IN TEXT AND FIGURE LEGENDS

Albian top stage of Lower Cretaceous
Arenig Series Lower Ordovician

Bembridge Limestone Oligocene
Bradford Clay Middle Jurassic

Cenomanian lowest stage of Upper Cretaceous
Chalk Upper Cretaceous
Chalk Marl Upper Cretaceous
Clypeus Grit Middle Jurassic
Coal Measures Upper Carboniferous
Corallian Upper Jurassic
Coralline Crag Pliocene
Crag deposits unconsolidated shallow water marine beds of Plio-Pleistocene age; see also Red and Coralline Crag

Dalradian Series rocks of late Precambrian and early Palaeozoic age
Deltaic Series Middle Jurassic
Downtonian lowest stage of the Old Red Sandstone
Durness Limestone Lower Cambrian and Lower Ordovician in Northwest Scotland

Faringdon Gravels Lower Cretaceous

Gault Lower Cretaceous
Great Oolite Middle Jurassic
Greensand Lower Cretaceous

Inferior Oolite Middle Jurassic

Kimmeridge Clay Upper Jurassic

Lias Lower Jurassic
London Clay Eocene
Ludlow Series Upper Silurian
Ludlow Bone Bed thin pebble bed with remanié vertebrate debris at the base of the Downtonian (q.v.)

Magnesian Limestone Upper Permian (Zechstein)
Millstone Grit Series Upper Carboniferous (Namurian)
Muschelkalk Middle Trias (in Germany)

Oxford Clay Upper Jurassic

Portland Limestone Upper Jurassic
Purbeck Beds Uppermost Jurassic and basal Cretaceous
Purbeck 'dirt' beds Uppermost Jurassic

Red Crag Lower Pleistocene
Rhaetic Upper Trias

Stonesfield Slate Middle Jurassic

Tremadoc Series Uppermost Cambrian (some authors include it in the Ordovician)

Wealden Beds Lower Cretaceous
Wenlock Limestone Middle Silurian

Index

Names of genera are italicised. Page numbers of illustrations are in bold type. For a simple definition of morphological terms, see also the glossary of technical terms listed under the appropriate groups. Subsidiary items entered under the main groups of fossils are in page order.